London Mathematical Society Lecture Note Series. 116

Representations of Algebras

Proceedings of the Durham symposium 1985

Edited by P. WEBB

Department of Mathematics, University of Manc

CAMBRIDGE UNIVERSITY PRESS

Cambridge

London New York New Rochelle

Melbourne Sydney

CAMBRIDGE UNIVERSITY PRESS
Cambridge, New York, Melbourne, Madrid, Cape Town, Singapore,
São Paulo, Delhi, Dubai, Tokyo, Mexico City

Cambridge University Press
The Edinburgh Building, Cambridge CB2 8RU, UK

Published in the United States of America by
Cambridge University Press, New York

www.cambridge.org
Information on this title: www.cambridge.org/9780521312882

First published 1986

A catalogue record for this publication is available from the British Library

Library of Congress cataloging in publication data applied for

ISBN 978-0-521-31288-2 Paperback

CONTENTS

INTRODUCTION

P.J. Webb
Department of Mathematics, The University, Manchester M13 9PL.

This volume presents the core of expository lectures given at the 1985 Durham symposium on representations of algebras. Between them they survey a good part of the cutting edge of research in the area of diagrammatic representation theory, and also with the article by Benson, some recent modular representation theory of groups. The lectures were given to a mixed audience of specialists in various parts of algebra, and it was not possible to assume that everyone was up to the minute with the jargon. An attempt has been made to write down the lectures in a similar spirit in the hope that their readership will be as wide as possible.

It is not easy to get into the area of 'quiver' representation theory from the outside because the area has been expanding so rapidly, and there is no single text where all the fundamentals are developed. Unfortunately the beginner has to read around a bit, and to help with this and set the scene for the articles in this book I propose to give a brief sketch of the way the subject has gone, indicating where accounts of the theory may be found. Useful general references are Gabriel (1980), and Ringel (1984), Chapter 2.

The subject began in the late 1960's with the work of Gabriel, Auslander and several Soviet mathematicians, including Nazarova and Roiter. Gabriel considered representations of quivers because these simultaneously deal with various types of matrix classification and also representations of algebras. See Section 1 of the article by Kraft and Riedtmann for the definitions, also Ringel (1984). Gabriel's theorem that the quivers with only finitely many isomorphism classes of indecomposable modules (= finite representation type) are precisely the Dynkin diagrams led to a conceptual approach in the beautiful paper of Bernstein et al. (1973), where it became evident that finiteness of representation type depends on the positive definiteness of an associated quadratic form, the Tits form (see Kraft and

Riedtmann, this volume), and this is precisely the criterion which distinguishes the Dynkin diagrams from other graphs. This approach also gave rise to a way of constructing all the indecomposable representations of the Dynkin diagrams by means of 'Coxeter functors' or 'reflection functors', so called because they act as reflections on an associated vector space equipped with the Tits form. At this stage the calculations take on a purely combinatorial flavour which leaves the algebra behind, and one has the impression of having got something for nothing. See Gabriel (1980) for these calculations. The direct generalisation of this theory is the identification of the extended Dynkin diagrams as the quivers of tame type and the rest as wild and for this the reader should turn to section 1 of the article by Kraft and Riedtmann. The remainder of that article describes the achievement of Kac in identifying the set of dimension vectors of indecomposable representation of an arbitrary quiver with the positive roots in the corresponding root system (in the sense of Kac-Moody). In reading this it is a great help to have some feel for the properties of the root systems as described in Chapter 5 of Kac (1985).

It is convenient to describe quivers before anything else because the basic notions are quite elementary, involving just linear algebra. But we are really interested in modules over rings, and in fact quiver representations are really the same as modules over an associated algebra, called the path algebra (see 2.1 of Ringel (1985) for this and what follows). This algebra is always hereditary and so is rather special, but an arbitrary finite dimensional algebra is always Morita equivalent to a quotient of a path algebra by an ideal (of relations). Such a set-up consisting of a quiver together with generators for an ideal in the path algebra is called a 'quiver with relations' and this is the approach which will frequently be taken in Ringel's article in this volume. It is a powerful way of describing an algebra and is a means by which the combinatorial methods of quivers are introduced into module categories.

The main link between the combinatorics of quivers and representations of algebras is by means of almost split sequences (which are also called Auslander-Reiten sequences). These came about through Auslander's determination of the simple objects in a certain functor category and the fact that they have resolutions by finitely generated projective functors in many situations (see Gabriel (1980)). At the time this approach was received badly, because it was hard to see the reason for having to go into functor categories. To remedy this situation Auslander and Reiten translated the statement about the existence of a resolution of functors to a statement about the existence of sequences of modules, whereupon it achieved a far more widespread success. It must be said, however, that although the simplest way to learn about this for the first time is via the sequences (as in Pierce (1982) for example), the most satisfactory conceptual approach is via the functors. Both are a good idea. In his first article in this volume Auslander surveys the state of the art of the circumstances in which almost split sequences exist. When they do exist, they are short exact sequences of modules $0 \to A \to B \to C \to 0$ which are uniquely specified once either of their end terms A or C is specified. We write $A = \tau C$ and call τ the 'Auslander translate'. Apart from the elegance of their definition, it is hard to see the reason for studying these sequences until certain connections are pointed out. Thus, in the case of path algebras of quivers the Auslander translate is practically the same as one of the Coxeter transformations considered by Bernstein, Gelfand, Ponomarev, the exact relation being given in section 5.4 of Gabriel (1980), and its significance demonstrated later in the same article by constructions of Auslander-Reiten quivers. It became apparent during the seventies that the Auslander-Reiten quiver has some rather simple combinatorial properties and at the same time contains considerable information about the module category, which in the case of finite representation type is more or less complete. There are now many achievements of this approach, and one might cite as examples the approach by Auslander to the first Brauer-Thrall conjecture (proved by Roiter, see Ringel (1980)), the classification by Riedtmann of self-injective algebras of finite representation type (in a series of papers), a variety of other classifications by various authors, the criterion for finite representation type according to Bongartz-Happel-Vossieck (see Ringel's article, these Proceedings), the theorem on the existence of a multiplicative basis in finite representation type algebras by Bautista et al..

A casual glance at pictures of quivers (for example, in this book) reveals that for the most part they seem to consist of regular-looking meshes of arrows which replicate themselves across the page. This is no accident, and depends on the fact that application of the Auslander translate preserves the quiver structure. The best way to view this is via the abstract notion of a translation quiver (Ringel's article), to which

Riedtmann's Structure Theorem applies (Riedtmann (1980) or Benson (1984), 2.29). It leads to the whole area of covering theory, for which there does not seem to be an adequate self-contained reference. This approach is implicit at points in Ringel's article, and in the work of many authors.

Whereas the Coxeter functors are only defined for hereditary algebras, the Auslander translate has the obvious advantage of working for all finite dimensional algebras. It was perhaps with this in mind that Brenner and Butler (1980) produced a formulation for all algebras of the reflection functors which appear en route in the Coxeter functor definition. They called their new functors tilting functors, and while one can read their original paper it is better to read an account which illustrates their use. For this, lecture 2 of Ringel's article, these Proceedings, will serve.

The article by Knörrer and the second article by Auslander in this book deal (amongst other things) with the representation theory of local rings of singularities. Much recent interest in this derives from the observation by McKay that when G is a finite subgroup of SL(2,C) a graph obtained by the decomposition of tensor products of the irreducible modules is an extension by one point of the desingularisation graph of the corresponding Kleinian singularity (which is defined as the fixed-point ring in the action of G on the power series ring in two variables). For this background, Slodowy (1983) is useful. Finally, in connection with the article of Roiter I would recommend as background reading the remarks of Ringel in lecture 1 of his article here for the connection between poset representations and quiver representations, and also section 2.6 of Ringel (1984). Kleiner's theorem on the posets of finite representation type was originally proved using the 'differentiation process' due to Nazarova and Roiter, and this proof is presented here.

REFERENCES

Bautista, R., Gabriel, P., Roiter, A.V., Salmeron, L. (to appear in Invent. Math.). Representation-finite algebras and multiplicative basis.

Benson, D. (1984). Modular Representation Theory: New Trends and Methods, Springer Lecture Notes in Mathematics 1081.

Bernstein, I.N., Gelfand, I.M., Ponomarev, V.A. (1973). Coxeter functors and Gabriel's theorem. Russian Math. Surveys 28, 17-32.

Brenner, S., Butler, M.C.R. (1980). Generalizations of the

Bernstein-Gelfand-Ponomarev reflection functors. Proc. ICRA II (Ottawa 1979), Springer Lecture Notes in Math. 832, 103-169.

Gabriel, P. (1980). Auslander-Reiten sequences and representation-finite algebras, Proc. ICRA II (Ottawa 1979), Springer Lecture Notes In Math. 831, 1-71.

Kac, V.G. (1985). Infinite Dimensional Lie Algebras. Cambridge Univ. Press.

Pierce, R.S. (1982). Associative algebras. Graduate Texts in Mathematics 88, Springer (New York, Berlin).

Riedtmann, C. (1980). Algebren, Darstellungsköcher, Ueberlagerungen und Zürück, Comm. Math. Helvetici 55, 199-224.

Ringel, C.M. (1980). Report on the Brauer-Thrall conjectures, Proc. ICRA II (Ottawa 1979), Springer Lecture Notes in Mathematics 831, 104-136.

Ringel, C.M. (1984). Tame Algebras and Integral Quadratic Forms. Springer Lecture Notes in Mathematics 1099.

Slodowy, P. (1983). Platonic solids, Kleinian singularities, and Lie groups Springer Lecture Notes in Mathematics 1008, 102-138.

REPRESENTATION THEORY OF FINITE-DIMENSIONAL ALGEBRAS

DURHAM LECTURES 1985

CLAUS MICHAEL RINGEL
Fakultat fur Mathematik, Universitat Bielefeld,
Universitatsstrasse, D-4800 Bielefeld-1, West Germany.

Let k be an algebraically closed field and A a finite-dimensional k-algebra (associative, with 1). We consider finite-dimensional left A-modules, and call them just modules; the category of all A-modules will be denoted by A-mod. Any module can be written as a (finite) direct sum of indecomposable modules, and the theorem of Krull-Schmidt asserts that such a decomposition is essentially unique: it is unique up to isomorphism. For many purposes it therefore is sufficient to deal only with indecomposable modules. The main problems of the representation theory of finite-dimensional algebras are the following:

- to develop methods for constructing indecomposable modules,
- to look for suitable invariants in order to be able to identify indecomposable modules,
- to show that a given list of indecomposable modules is complete: that it contains a representative of any isomorphism class.

Typical invariants of a module M are the socalled Jordan-Hölder multiplicities: the algebra A has only finitely many simple modules, say $E_1,\dots,E_n$, and we may denote by $(\underline{\dim}\, M)_i$ the multiplicity of E_i occurring in a composition series of M (this is well-known to be an invariant of the isomorphism class of M). The vector $\underline{\dim}\, M$ obtained in this way is called the dimension vector of M. So one may ask for a description of the possible dimension vectors of indecomposable modules for a given algebra, and, having fixed a particular dimension vector, for a description of all indecomposable modules having this dimension vector.

One of the first questions usually will be that about the number of isomorphism classes of indecomposable modules. There may be only finitely many isomorphism classes of indecomposable A-modules, and then A is said to be representation-finite. Examples of representation-finite algebras are first of all the semi-simple ones, but also the algebras of all upper triangular matrices of given size, and there is a vast literature on representation-finite algebras. In case there are infinitely many isomorphism classes of indecomposable A-modules, there are actually always one-parameter families of isomorphism classes of indecomposable A-modules, as was conjectured by Brauer and Thrall. If there exists a two-parameter family of isomorphism classes of indecomposable A-modules, A is said to be wild, otherwise tame. The study of representation-infinite algebras is still in the beginning, only some types of examples seem to be well understood. We will present below several results which are independent of the representation type and exhibit some examples of tame algebras. In addition, we will pose a number of open problems which seem to be worthwhile to study.

A general reference for the terminology used here are our lecture notes [Ri2]. Those notes should also be consulted for the precise attribution of most of the results presented here. Only in case we deal with results which fell out of the scope of [Ri2] or which where not yet available at that time, we will indicate the source. Our aim in these lecturesis to give an introduction to the representation theory of finite-dimensional algebras. In particular, we are going to direct the interest towards the main results presented in [Ri2]. In addition, we will report on some recent investigations which are contained in the papers [RV], [Ri3], [Ha] and [HR].

LECTURE 1

THE AUSLANDER-REITEN QUIVER

It will be necessary to consider besides categories of the form A-mod also some related categories, for example full subcategories of A-mod (which are closed under direct sums and direct summands), or the categories of representations of partially ordered sets, or derived categories. Always, the categories which we will deal with will be k-additive categories (thus, additive categories with k operating centrally on the Hom-sets and such that all $\mathrm{Hom}(X,Y)$ are finite-dimensional k-vector-spaces) with split idempotents; and we call such a category a Krull-Schmidt category (note that in a Krull-Schmidt category, any object is a finite direct sum of indecomposable objects, and such a decomposition is unique up to isomorphism).

We start with the basic notions. Given an indecomposable object in a Krull-Schmidt category we call a map $f : X \longrightarrow Y$ a source map for X (the usual name would be "minimal left almost split map") provided the following three conditions are satisfied: first, f is not split mono; second, given any map $f' : X \longrightarrow Y'$ which is not split mono, there is $\eta : Y \longrightarrow Y'$ with $f' = f\eta$; and third, any $\zeta : Y \longrightarrow Y$ with $f = f\zeta$ is an automorphism. There is the following dual notion: Given an indecomposable module Z, we call a map $g : Y \longrightarrow Z$ a sink map for Z (or a "minimal right almost split map") provided, first, g is not split epi; second, given any map $g' : Y' \longrightarrow Z$ which is not split epi, there is η with $g' = \eta g$; and third, any $\zeta: Y \longrightarrow Y$ with $g = \zeta g$ is an automorphism. In case we deal with $\mathcal{K}$ = A-mod where A is a representation-finite algebra, it is not surprising to see that source maps and sink maps exist. The following remarkable theorem asserts that they always do exist in module categories, independent of the representation type:

THEOREM (M. Auslander, I. Reiten). Let A be a finite-dimensional k-algebra. For any indecomposable module M, there exists a source map and a sink map in A-mod, and both are unique up to isomorphism.

Let Z be indecomposable with sink map $g : Y \longrightarrow Z$. Either Z is projective, then we may take for Y the radical $\mathrm{rad}\, Z$ of Z and for g the inclusion map. Or, if Z is not projective, then g is epi, its kernel $\mathrm{Ker}\, g$ is indecomposable and will be denoted by τZ, and the inclusion map $\tau Z \longrightarrow Y$ is a source map.

Dually, let X' be indecomposable with source map $f' : X' \longrightarrow Y'$. Either X' is injective, then we may take $Y' = X'/\operatorname{soc} X'$, and f' the canonical epimorphism. Or, if X' is not injective, then f' is mono, its cokernel $\operatorname{Cok} f'$ is indecomposable and will be denoted by $\tau^{-}X'$, and the canonical epimorphism $Y' \longrightarrow \tau^{-}X'$ is a sink map.

Starting with a non-projective indecomposable module Z, or with a non-injective indecomposable module X, we obtain a non-split exact sequence

$$0 \longrightarrow X \xrightarrow{f} Y \xrightarrow{g} Z \longrightarrow 0$$

with f a source map for X, and g a sink map for Z, with $X = \tau Z$ and $Z = \tau^{-}X$. Sequences of this kind are called Auslander-Reiten sequences. Now, in such an Auslander-Reiten sequence, both modules X, and Z are indecomposable, whereas Y usually is not. We decompose $Y = \bigoplus_i Y_i$, with all Y_i indecomposable, and rewrite the sequence above in the form

$$(*) \qquad 0 \longrightarrow X \xrightarrow{(f_i)_i} \bigoplus_i Y_i \xrightarrow{(g_i)_i} Z \longrightarrow 0 .$$

The maps $f_i : X \longrightarrow Y_i$ are irreducible (we recall the definition below), and we obtain in this way sufficiently many irreducible maps starting in X. The maps $g_i : Y_i \longrightarrow Z$ also are irreducible, and we obtain in this way sufficiently many irreducible maps ending in Z.

Consider a general Krull-Schmidt category $\mathcal{K}$. If M, M' are indecomposable objects in $\mathcal{K}$, denote by $\operatorname{rad}(M,M')$ the set of non-invertible maps $M \longrightarrow M'$. If M, M' are arbitrary, say with decompositions $M = \bigoplus M_i$, $M' = \bigoplus M'_j$ into indecomposables, let $\operatorname{rad}(M,M') = \bigoplus_{i,j} \operatorname{rad}(M_i, M'_j)$. We obtain in this way an ideal rad in the category $\mathcal{K}$. We define $\operatorname{rad}^d(M,M')$ as the set of maps $M \longrightarrow M'$ which can be written as compositions of d maps all belonging to rad, and let $\operatorname{rad}^\infty = \bigcap_{d \in \mathbb{N}} \operatorname{rad}^d$. If M, M' are indecomposable objects, the maps in $\operatorname{rad}(M,M') \setminus \operatorname{rad}^2(M,M')$ are just the irreducible maps, and the factorspace $\operatorname{Irr}(M,M') = \operatorname{rad}(M,M')/\operatorname{rad}^2(M,M')$ is called the bimodule of irreducible maps.

We have noted above that in the module category A-mod, an Auslander-Reiten sequence $(*)$ displays sufficiently many irreducible maps starting in X or ending in Z. In fact, assume that M is indecomposable, and that $\operatorname{Irr}(X,M)$ is of dimension d. Then, precisely d of the summands Y_i of Y are isomorphic to M, say $Y_1 = \ldots = Y_d = M$, and the residue classes of the maps $f_1,\dots,f_d$ form a basis of $\mathrm{Irr}(X,M)$, whereas the residue classes of the maps $g_1,\dots,g_d$ form a basis of $\mathrm{Irr}(M,Z)$. In particular, we always have

$$(**)\qquad \dim_k \mathrm{Irr}(\tau Z,M) = \dim_k \mathrm{Irr}(M,Z)$$

for M,Z indecomposable, and Z not projective.

With these preparations, we are going to define the Auslander-Reiten quiver Γ_A of A. We have noted in the introduction that the main object of the representation theory is the study of the set of isomorphism classes of indecomposable modules, and we denote this set by $(\Gamma_A)_o$. We want to endow this set with more structure in order to gain insight into its properties, and the theory expounded above shows that we may consider it as the set of vertices of a socalled translation quiver. Now, a quiver is nothing else than an oriented graph with possible multiple arrows and loops, thus of the form $Q = (Q_o,Q_1,s,e)$, where Q_o,Q_1 are two sets, and $s,e : Q_1 \longrightarrow Q_o$ set maps; the elements of Q_o are called vertices or points, those of Q_1 arrows, and given $\alpha \in Q_1$, then $s(\alpha)$ is called its starting point and $e(\alpha)$ its end point, pictured as follows: $s(\alpha) \xrightarrow{\alpha} e(\alpha)$. A translation quiver Γ is a locally finite quiver with an additional bijection $\tau = \tau_\Gamma : \Gamma'_o \longrightarrow \Gamma''_o$ of two subsets of Γ_o such that for $y \in \Gamma_o$, $z \in \Gamma'_o$, the number of arrows from y to z coincides with the number of arrows from τz to y. (In case we actually fix bijections σ from the set of arrows $y \longrightarrow z$ onto the set of arrows $\tau z \longrightarrow y$, we speak of a polarized translation quiver). A translation quiver may be thought of as being built from small units, the socalled "meshes"

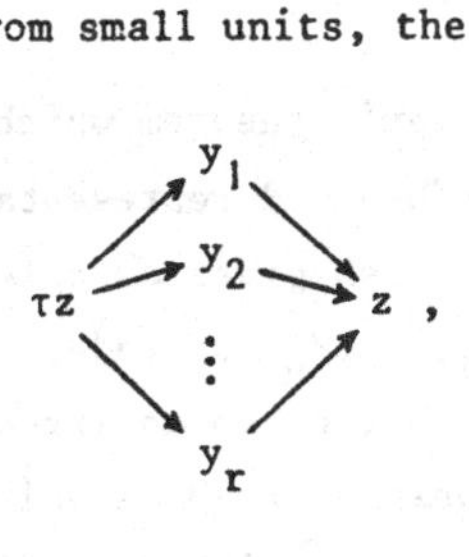

they are defined for any $z \in \Gamma'_o$ (some of the y_i may coincide). The vertices in $\Gamma_o \setminus \Gamma'_o$ are called projective vertices, those in $\Gamma_o \setminus \Gamma''_o$ are called injective vertices. We return to the case of a finite-dimensional algebra A. The isomorphism class of a module M will be denoted by $[M]$. We have already noted that the vertices of Γ_A are of the form

[X], with X an indecomposable module. For X,Y indecomposable, the number of arrows $[X] \longrightarrow [Y]$ in Γ_A is given by $\dim_k \operatorname{Irr}(X,Y)$, thus there is at least one arrow $[X] \to [Y]$ iff there exists an irreducible map $X \longrightarrow Y$. As translation we use the function τ, with $\tau[Z] = [\tau Z]$ which is defined for Z indecomposable and non projective. The formula (**) asserts that indeed we obtain in this way a translation quiver, the <u>Auslander-Reiten quiver</u> Γ_A of A. Of course, the projective vertices of Γ_A are just the vertices of the form [P], with P indecomposable projective, the injective vertices are those of the form [Q] with Q indecomposable injective. We observe that Γ_A has only finitely many projective and finitely many injective vertices, and these numbers coincide.

The structure of the Auslander-Reiten quiver Γ_A of finite-dimensional algebras should be studied rather carefully. Usually, Γ_A will decompose into several components and one may ask for the possible translation quivers occurring as components of Auslander-Reiten quivers. If X is an indecomposable module, and Γ a component of Γ_A, containing [X], we just will say that X belongs to Γ. The components which do not contain projective or injective vertices will be said to be <u>regular</u>; of course, all but finitely many components are regular. Since Γ_A is always locally finite, any component is either finite or countable. In fact, for representation-infinite (and connected) algebras, there are no finite components:

<u>THEOREM</u> (Auslander). Assume A is connected, and that Γ is a component of Γ_A containing only modules of bounded length. Then A is representation finite and $\Gamma_A = \Gamma$.

This strengthens Rojter's theorem which had established the first Brauer-Thrall conjecture ("bounded representation type implies finite representation type"). It also gives an effective method for showing that a given finite list $M_1,\ldots,M_n$ of indecomposable modules is complete: it is sufficient to show that given an irreducible map $M_i \longrightarrow X$ or $X \longrightarrow M_i$ with X indecomposable, then X already occurs in the list. The completeness may be shown in the following way: first, one checks that all indecomposable projective modules occur in the list; second, that the list is closed under τ^- and contains all indecomposable summands of the corresponding Auslander-Reiten sequences (i.e. if M_j is not injective, and (**) is an Auslander-Reiten sequence with $X = M_j$, then Z and all

the Y_i should belong to the list). We observe that this procedure refrains from the necessity of decomposing large modules, an operation which would produce a lot of difficulties. The only modules to be decomposed are those occurring as middle terms of Auslander-Reiten sequences (and, according to a theorem of Bautista-Brenner for representation-finite algebras, these are direct sums of at most four indecomposable modules).
We pose the following problem:

Problem 1. Let Γ be a component of some Auslander-Reiten quiver, and d a natural number. Is the number of isomorphism classes of indecomposable modules X in Γ of length d always finite?

Given any quiver Δ, one may construct a translation quiver $\mathbb{Z}\Delta$ as follows: the set of vertices of $\mathbb{Z}\Delta$ is given by $\mathbb{Z}\times\Delta_0$; for any arrow $\alpha : a \longrightarrow b$ in Δ, there are the arrows $(z,\alpha) : (z,a) \longrightarrow (z,b)$ and $(z,\alpha)' : (z,b) \longrightarrow (z+1,a)$ for all $z \in \mathbb{Z}$, and the translation is defined by $\tau(z,a) = (z-1,a)$. This is a regular translation quiver, and any regular translation quiver can be obtained from some $\mathbb{Z}\Delta$, even with Δ a tree, as factor quiver $\mathbb{Z}\Delta/G$ with respect to some automorphism group G of $\mathbb{Z}\Delta$. Note that for Δ an oriented tree, we obtain isomorphic translation quivers when changing the orientation, thus, in this case, we do not have to specify the orientation. Of particular interest for representation theory are the translation quivers $\mathbb{Z}\mathbb{A}_\infty$, $\mathbb{Z}\mathbb{A}_\infty^\infty$ and $\mathbb{Z}\mathbb{D}_\infty$, where $\mathbb{A}_\infty^\infty$ has as vertices the integers, and edges i—$i+1$, for $i \in \mathbb{Z}$, the graph $\mathbb{A}_\infty$ is the full subgraph of $\mathbb{A}_\infty^\infty$ given by the non-negative integers, and $\mathbb{D}_\infty$ is obtained from $\mathbb{A}_\infty$ by adding a vertex $0'$ and an edge $0'$—1. Thus,

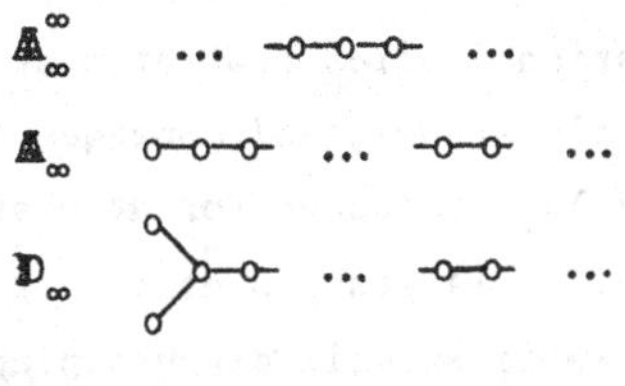

The automorphisms of $\mathbb{Z}\mathbb{A}_\infty$ are of the form τ^n, with $n \in \mathbb{Z}$, and for $n \geq 1$ we denote $\mathbb{Z}\mathbb{A}_\infty/\langle\tau^n\rangle$ just by $\mathbb{Z}\mathbb{A}_\infty/n$. Components of the form $\mathbb{Z}\mathbb{A}_\infty/n$ with $n \geq 1$ will be called regular tubes, those with $n = 1$ are said to be homogeneous tubes.

The regular components of a hereditary algebra A are all of the form $\mathbb{Z}\mathbb{A}_\infty$ or $\mathbb{Z}\mathbb{A}_\infty/n$, with $n \geq 1$, they are tubes in case A is tame, and then almost all are homogeneous, whereas all regular components are of the form $\mathbb{Z}\mathbb{A}_\infty$, in case A is wild [Ri1]. For the group algebra kG of a dihedral 2-group G, with k of characteristic 2, there are countably many components of the form $\mathbb{Z}\mathbb{A}_\infty^\infty$, all other regular components are homogeneous tubes, the number of homogeneous tubes is equal to the cardinality of k [BSh]. For a semidihedral group G, and k again of characteristic 2, there are countably many components of the form $\mathbb{Z}\mathbb{A}_\infty^\infty$, of the form $\mathbb{Z}\mathbb{D}_\infty$, and of the form $\mathbb{Z}\mathbb{A}_\infty/2$; the remaining ones are homogeneous tubes, the number of such components again is equal to the cardinality of k. We will see in the next lecture that for nearly all finite quivers Δ, there are algebras having a component of the form $\mathbb{Z}\Delta$.

A vertex x of a translation quiver which satisfies $\tau^n x = x$ for some $n \geq 1$ is said to be <u>periodic</u>. Of course, all vertices of $\mathbb{Z}\mathbb{A}_\infty/n$ are periodic. And there is the following converse

<u>PROPOSITION</u>. If a regular component of an Auslander-Reiten quiver contains a periodic vertex, then it is a regular tube.

We pose the following problem:

<u>Problem 2</u>. Let A be any finite-dimensional algebra. Is it true that all but finitely many components of Γ_A are of the form $\mathbb{Z}\mathbb{A}_\infty$, $\mathbb{Z}\mathbb{A}_\infty/n$, $\mathbb{Z}\mathbb{A}_\infty^\infty$, and $\mathbb{Z}\mathbb{D}_\infty$? Is it true that all but at most countably many components of Γ_A are of the form $\mathbb{Z}\mathbb{A}_\infty$ and $\mathbb{Z}\mathbb{A}_\infty/1$?

We note that the first question has a positive answer for group algebras, according to the investigations of Webb [W]. Note that a positive answer to the first question implies that at most finitely many components of Γ_A can have multiple arrows, and consequently that for M,M' indecomposable, $\dim \mathrm{Irr}(M,M') \leq 1$, except for at most countably many modules M,M'. On the other hand, it is easy to construct examples of components with multiple arrows: We later will discuss preprojective components, and there is no difficulty to exhibit such components with an arbitrarily large number of arrows between two vertices. In the next lecture, we will see that in the same way we also may construct regular components with an arbitrarily large number of arrows between two vertices.

Given a polarized translation quiver Γ, there is defined its <u>mesh category</u> $k(\Gamma)$, as follows: first, we construct the path category $k\Gamma$

of Γ, it is an additive category whose indecomposable objects are just the vertices of Γ, the space $\mathrm{Hom}_{k\Gamma}(a,b)$ of morphisms from a to b is the k-vectorspace with basis the set of all paths from a to b in Γ, and the composition of morphisms is induced from the usual composition of paths. The mesh category $k(\Gamma)$ is the factor category of $k\Gamma$ modulo the ideal generated by the socalled mesh relations: these are the elements of the form $\Sigma\, \sigma(\alpha)\alpha$, the summation being done over all arrows α with fixed non-projective endpoint.

A component Γ of the Auslander-Reiten quiver Γ_A of an algebra A will be said to be <u>standard</u> provided the full additive subcategory of A-mod whose indecomposable objects are the modules in Γ is equivalent to $k(\Gamma)$. In this and the next lecture, we will provide methods for constructing standard components, but we should stress already here that standard components seem to occur only scarcely.

<u>Problem 3</u>. Let A be any finite-dimensional algebra. Is it true that any standard regular component is either a regular tube or of the form $\mathbb{Z}\Delta$, with Δ a finite quiver without oriented cycles?

Given a translation quiver Γ, it will be of interest to consider integer valued functions on the set Γ_0 of vertices of Γ. A function $f : \Gamma_0 \longrightarrow \mathbb{Z}$ is called an <u>additive</u> function on Γ provided

$$f(z) + f(\tau z) = \sum_{e(\alpha)=z} f(s(\alpha))$$

for any non-projective vertex z. A typical example of an additive function on an Auslander-Reiten quiver Γ_A is the length function which attaches to each vertex $[X]$ the length of the module X. This function has additional properties, giving rise to the following definition: an additive function f on the translation quiver Γ is called a <u>length function</u> provided

$$f(x) \geq 1 \qquad \text{for all } x \in \Gamma_0$$

$$f(p) = 1 + \sum_{e(\alpha)=p} f(s(\alpha)) \text{ for any projective vertex } p,$$

$$f(q) = 1 + \sum_{s(\alpha)=q} f(e(\alpha)) \text{ for any injective vertex } q.$$

We note however that on an Auslander-Reiten quiver Γ_A, there usually will exist length functions different from the function $[X] \mapsto \mathrm{length}(X)$.

A translation quiver Γ is said to be preprojective provided the following three conditions are satisfied:

(1) There is no cyclic path in Γ.

(2) Any τ-orbit contains a projective vertex.

(3) There are only finitely many τ-orbits.

There is the following criterion for preprojective translation quivers occurring as components of Auslander-Reiten quivers of algebras:

PROPOSITION. A prepojective translation quiver admits at most one length function. A connected preprojective translation quiver occurs as a component of some Auslander-Reiten quiver Γ_A if and only if it admits a length function.

Also, there is the following result:

PROPOSITION. A preprojective component of an Auslander-Reiten quiver Γ_A always is standard.

Given a Krull-Schmidt category K, we construct full subcategories ${}_dK$, for d an integer ≥ -1, or $d = \infty$, as follows: Let ${}_{-1}K$ be the subcategory $\langle o \rangle$. If ${}_{d-1}K$ is already defined, let ${}_dK$ be the full subcategory of all objects Z of K which have the property that any indecomposable object Y with $\mathrm{rad}(Y,Z) \neq 0$ belongs to ${}_{d-1}K$. Finally, let ${}_\infty K$ be the union of all ${}_dK$, $d \in \mathbb{N}$. In case K = A-mod, the indecomposable modules in ${}_0K$ are just the simple projective modules; and an indecomposable module belongs to ${}_1K$ if and only if it is projective and its radical is semisimple and projective. Clearly, if $[X] \longrightarrow [Y]$ is an arrow in Γ_A, and Y belongs to ${}_\infty$(A-mod), then also X belongs to ${}_\infty$(A-mod). Actually, given a translation quiver Γ, we may define in a similar way full translation subquivers ${}_d\Gamma$, where d is an integer ≥ -1, or $d = \infty$ (by definition, ${}_{-1}\Gamma$ is the empty quiver, a vertex z of Γ belongs to ${}_d\Gamma$ if and only if every vertex y with an arrow $y \longrightarrow z$ belongs to ${}_{d-1}\Gamma$, and ${}_\infty\Gamma$ is the union of all ${}_d\Gamma$, $d \in \mathbb{N}$), and it is easy to see that a vertex [X] of Γ_A belongs to ${}_d\Gamma_A$ if and only if X belongs to ${}_d$(A-mod), for any d. We are interested in the question under what conditions ${}_\infty\Gamma_A$ is a component, or at least a union of components of Γ_A. Note that ${}_\infty\Gamma_A$ is non-empty if and only if there exists a simple projective A-module. Always ${}_\infty\Gamma_A$ is a preprojective translation quiver, and any component of Γ_A which is preprojective, is contained in ${}_\infty\Gamma_A$. The following criterion is easily verified: ${}_\infty\Gamma_A$ is a union of (preprojective) components if and only if the following condition is satisfied:

(P) If P is an indecomposable projective A-module, and some indecomposable direct summand of rad P belongs to ${}_\infty$(A-mod), then rad P belongs to ${}_\infty$(A-mod).

An immediate consequence is the following: assume the radical of any indecomposable projective A-module is indecomposable or zero, and that there exists at least one simple projective A-module, then ${}_\infty\Gamma_A$ is non-empty and a union of preprojective components of Γ_A.

Examples of classes of algebras which have preprojective components are the following:

- the hereditary algebras
- the concealed algebras (see lecture 2)
- the canonical algebras (see lecture 3).

One should observe that the Auslander-Reiten quiver of an algebra A usually will not have any preprojective component at all, but even for A being connected, there may be more than one preprojective component, as the following example shows:

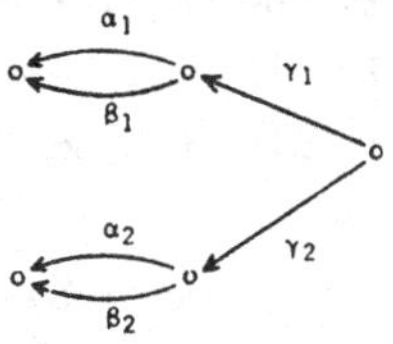

$$\gamma_1\alpha_1 = 0, \qquad \gamma_2\alpha_2 = 0$$

(Here, and in the following, it often will be convenient to exhibit an algebra A by a quiver with relations, this means that the opposite algebra A^{op} is obtained from the path algebra of the quiver by factoring out the ideal generated by the given relations. In this way, the category A-mod is just the category of all representations of the quiver which satisfy the given relations! We should recall that a representation V of a quiver Q is given by a set of (finite-dimensional) vectorspaces V_x, indexed by the vertices x of Q, and linear maps $V_\alpha : V_{s(\alpha)} \longrightarrow V_{e(\alpha)}$, indexed by the arrows α of Q.)

Let us outline the actual construction of a preprojective component by considering the following example:

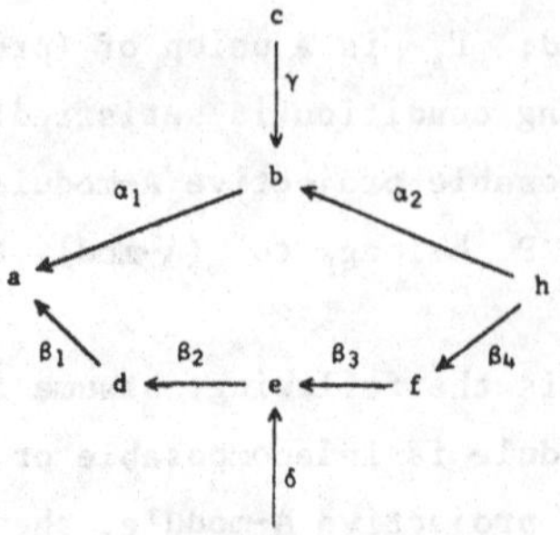

$$\alpha_2\alpha_1 = \beta_4\beta_3\beta_2\beta_1$$

First, one has to determine the indecomposable projective modules and their radicals. Any vertex x of the quiver Q of A determines a simple module E(x) (if we consider E(x) as a representation of Q, the vectorspace indexed by x is k, all others are 0), and we denote by P(x) the projective cover of x (considering P(x) as a representation of Q, the vectorspace $P(x)_y$ is obtained from the free vectorspace with basis all paths from x to y as a factorspace by taking into account the given relations). In this way, we obtain all simple and all indecomposable projective modules. Note that for any representation V of Q, the dimension of V_x is just the Jordan-Hölder multiplicity of E(x) in V, and it seems to be rather suggestive to arrange the entries of the dimension vector in the form of the quiver. Actually, it will be convenient to denote V just by dim V, provided V is the only indecomposable module with this dimension vector. (The indecomposable modules in $_\infty$(A-mod) always have this property, an account of this result will be given in the second lecture). In our case, we obtain the following list of indecomposable projective modules and their radicals:

x	a	b	c	d	e	f	g	h
P(x)	o o 1 ooo o o	o 1 1 ooo o o	1 1 1 ooo o o	o o 1 1oo o o	o o 1 11o o o	o o 1 111 o o	o o 1 11o o 1	o 1 1 111 1 o
rad P(x)	0	P(a)	P(b)	P(a)	P(d)	P(e)	P(e)	o 1 1 111 o o

The module P(a) is simple projective, and all rad P(x), x ≠ a, are indesomposable, thus the criterion mentioned above asserts that A-mod has a preprojective component.

We are going to construct inductively ${}_d$(A-mod) and ${}_d\Gamma_A$. The indecomposable modules in ${}_0$(A-mod) are the simple projective ones. (In our case, ${}_0\Gamma_A$ consists of the single vertex [P(a)]. In particular, ${}_\infty\Gamma_A$ will be connected.) Suppose we have already constructed ${}_d$(A-mod), and ${}_d\Gamma_A$, for some $d \geq 0$. We single out the indecomposable modules X in ${}_d$(A-mod) which satisfy the following properties:

i) $x = [X]$ is an injective vertex of ${}_d\Gamma_A$, and there is an arrow $x \longrightarrow y$ in ${}_d\Gamma_A$, with y not belonging to ${}_{d-1}\Gamma_A$.

ii) If M is an indecomposable module in ${}_d$(A-mod) with an arrow $[M] \longrightarrow x$ in ${}_d\Gamma_A$, then either M is an injective A-module, or else [M] is not injective in ${}_d\Gamma_A$.

iii) If X is a direct summand of rad P, with P indecomposable projective, then P belongs to ${}_d$(A-mod).

Suppose X satisfies i), ii), iii). Consider the vector

$$- \underline{\dim}\, X + \sum_\alpha \underline{\dim}\, Y_\alpha ,$$

where the summation extends over all arrows $\alpha : [M] \longrightarrow [Y_\alpha]$ starting in [X]. In case this vector is not positive, X is injective, thus the τ-orbit of [X] in Γ_A ends in [X]. Or else, this vector is the dimension vector $\underline{\underline{\dim}}\, \tau^- X$, and $\tau^- X$ belongs to ${}_{d+1}$(A-mod); in this case, we obtain $[\tau^- X]$ as a vertex of ${}_{d+1}\Gamma_A$ outside ${}_d\Gamma_A$, with $\tau[\tau^- X] = [X]$ (the arrows ending in $[\tau^- X]$ are uniquely determined by the condition on ${}_{d+1}\Gamma_A$ to be a translation quiver). In this way, we will obtain several new vertices in ${}_{d+1}\Gamma_A$ all of which are non-projective. In case there exist indecomposable projective modules P not belonging to ${}_d$(A-mod), we have to check whether there exists such a P with rad P in ${}_d$(A-mod). Thus assume P is indecomposable projective, does not belong to ${}_d$(A-mod), rad $P = \bigoplus Y_i^{n_i}$ with all Y_i indecomposable, and pairwise non-isomorphic, and all Y_i in ${}_d$(A-mod). In this case, [P] is a vertex of ${}_{d+1}\Gamma_A$ outside ${}_d\Gamma_A$, and there are n_i arrows $[Y_i] \longrightarrow [P]$. This finishes the construction of ${}_{d+1}\Gamma_A$. Let us consider our example: suppose we have already constructed ${}_5\Gamma_A$, namely the part depicted below by solid arrows, and we are going to construct ${}_6\Gamma_A$. There are three indecomposable modules X which satisfy the conditions i), ii), iii), namely those with dimension vectors $\begin{smallmatrix} & 0 & \\ & 0 & \\ 0 & 100 & 0 \\ & 0 & \end{smallmatrix}$, $\begin{smallmatrix} & 1 & \\ & 1 & \\ 1 & 110 & 0 \\ & 0 & \end{smallmatrix}$, and $\begin{smallmatrix} & 0 & \\ & 1 & \\ 2 & 221 & 0 \\ & 1 & \end{smallmatrix}$, and the calculation of dimension vectors shows that none of these is injective (of course, this could be checked also directly!). In this way, we obtain three non-projective vertices in ${}_6\Gamma_A$ outside ${}_5\Gamma_A$. Also, the indecomposable module Y with

$\underline{\dim}\, Y = \begin{smallmatrix} & 0 & \\ & 1 & \\ 1 & 111 & 0 \\ & 0 & \end{smallmatrix}$ is just rad P(h), thus ${}_6\Gamma_A$ contains in addition [P(h)], and there is just one arrow ending in [P(h)], and this arrow starts in [Y]:

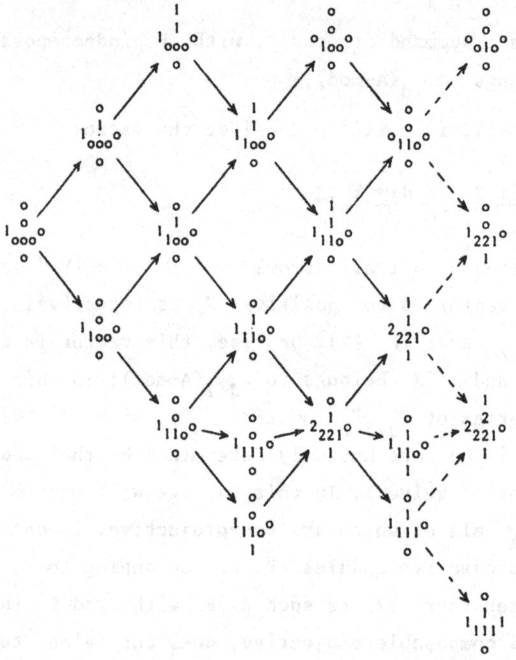

Continuing in this way, we obtain the complete component. In our case, the component turns out to be finite, thus it is the whole Auslander-Reiten quiver Γ_A. Actually, as soon as we know the position of the projective vertices, we may as well work solely with the length function: it only remains to determine whether any vertex is an injective one, and this can be read off from the values of the length function. In our case, we obtain the following values of the length function:

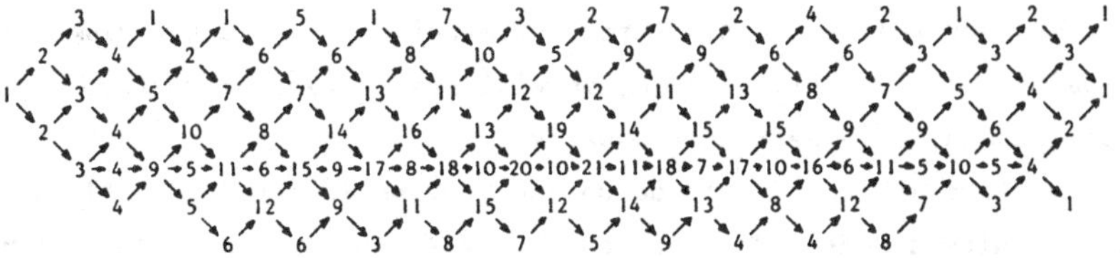

Looking at the Auslander-Reiten quiver Γ_A of an algebra A, the indecomposable A-modules seem to be converted into mere vertices of a graph. So, an obvious question is under what conditions one may recover the modules from the combinatorial data. In case we deal with a preprojective component Γ, everything works out very well: Let X be an indecomposable module. We decompose ${}_AA = \oplus P_i$, with P_i indecomposable (and projective), and A may be identified with the endomorphism ring $\mathrm{End}({}_AA) = \mathrm{End}(\oplus P_i)$, thus A is given by the various $\mathrm{Hom}(P_i,P_j)$ and their compositions. As a k-space, there are the canonical isomorphisms

$$X \approx \mathrm{Hom}({}_AA,X) \approx \oplus \mathrm{Hom}(P_i,X),$$

and the A-module structure on X corresponds to the operation of $\oplus \mathrm{Hom}(P_i,P_j)$ on $\oplus \mathrm{Hom}(P_i,X)$, using the composition of maps. Assume now that X belongs to Γ. Then, any P_i with $\mathrm{Hom}(P_i,X) \neq 0$ also belongs to Γ, since Γ is a preprojective component. Also, Γ is standard, thus we may calculate in $k(\Gamma)$ all the non-zero $\mathrm{Hom}(P_i,X)$, all $\mathrm{Hom}(P_s,P_t)$ where both P_s,P_t belong to Γ, and all the corresponding compositions. We may reformulate this as follows: Let e be an idempotent of A such that all the indecomposable direct summands of Ae belong to Γ, whereas none of $A(1-e)$ belongs to Γ. Then, first of all, $(1-e)Ae = o$, therefore $A(1-e)$ is a twosided ideal, and the algebra $A/A(1-e) \approx eAe$ can be calculated in $k(\Gamma)$. Also, given an indecomposable module X in Γ, we have $(1-e)X = o$, thus X as an A-module is in fact an $A/A(1-e)$-module, and X, as an $A/A(1-e)$-module, can be calculated in $k(\Gamma)$. Let us return to the example considered above. We have calculated the values of the length function, and we see that there is a unique indecomposable module of maximal length, namely $\tau^{-7}P(e)$. The explicit calculation of dimension vectors

shows that $\underline{\dim}\,\tau^{-7}P(e) = 2\begin{smallmatrix}&2&\\&3&\\3&4&3\\&2&\end{smallmatrix}2$. As a quiver representation, this module
is given by four 2-dimensional, three 3-dimensional, and one 4-dimensional
vectorspace. Later in this lecture, we will draw attention to these vectorspaces and suitably defined subspaces.

Translation quivers do not only arise in the representation
theory of finite-dimensional algebras, but also for many related structures. We will consider in this lecture a second case: the Auslander-Reiten quiver of a finite partially ordered set. First, we recall the main
notions of the representation theory of partially ordered sets. Let S
be a finite partially ordered set. An S-space is of the form
$V = (V_\omega; V_s)_{s \in S}$, where V_ω is a vectorspace, all V_s, $s \in S$, are subspaces of V_ω, and $V_s \subseteq V_t$ for $s \leq t$. The space V_ω is called the total
space of V, its dimension is denoted by $\underline{\dim}_\omega V$. A map $f : V \longrightarrow W$ of
S-spaces is given by a k-linear map $f_\omega : V_\omega \longrightarrow W_\omega$ satisfying
$V_s f_\omega \subseteq W_s$, for all $s \in S$; the restriction of f_ω to V_s will be denoted
by $f_s : V_s \longrightarrow W_s$ and we may write $f = (f_\omega; f_s)$. It will be convenient to
consider besides S also the partially ordered set S^+ obtained from S
by adding a new element ω with $s < \omega$ for all $s \in S$. We denote by
$\ell(S)$ the category of all S-spaces with finite-dimensional total space. It
is a Krull-Schmidt category with short exact sequences, a short exact se-
quence is of the form (f,g), where $f : V' \longrightarrow V$, $g : V \longrightarrow V''$ are maps
of S-spaces such that all the sequences

$$0 \longrightarrow V'_t \xrightarrow{f_t} V_t \xrightarrow{g_t} V''_t \longrightarrow 0$$

are exact, for $t \in S^+$. If (f,g) is a short exact sequence, then f may
be called a proper mono, and g a proper epi. Note that $f : V' \longrightarrow V$
is proper mono if and only if f_ω is mono and $V'_s f_s = V'_\omega f \cap V_s$. And,
$g : V \longrightarrow V''$ is proper epi if and only if all g_s, $s \in S^+$, are epi. For
$s \in S^+$, we define an S-space $P_S(s)$ by $(P_S(s))_t = k$ for all $t \geq s$,
and $= 0$ otherwise. These S-spaces behave similar to the indecomposable
projective modules: they are the only indecomposable S-spaces which have
the usual lifting property with respect to all proper epis. Similarly, we
may consider the indecomposable S-spaces which have the extension property
with respect to the proper monos: they are denoted by $Q_S(s)$ with
$s \in S \cup \{\omega'\}$. Here, $(Q_S(\omega'))_t = k$ for all $t \in S^+$, and, for $s \in S$, we

have $(Q_S(s))_t = k$ provided $t \nleq s$ and $= 0$ otherwise. Since $\ell(S)$ is a Krull-Schmidt category, we may speak of source maps and sink maps in $\ell(S)$, and there is the corresponding result which seems to be due to Bautista:

THEOREM. For any indecomposable S-space V, there exists a source map and a sink map, and both are unique up to isomorphism.

Let Z be an indecomposable S-space with sink map $g : Y \longrightarrow Z$. Either $Z = P_S(\omega)$, then $Y = 0$. Or, $Z = P_S(s)$ for some $s \in S$, then Y has one-dimensional total space; in particular, Y is indecomposable. Or, if Z is not of the form $P_S(s)$ for any $s \in S^+$, then there is an exact sequence (f,g) of S-spaces, say with $f : X \longrightarrow Y$, the S-space X is indecomposable and f is a source map.

Let X' be an indecomposable S-space with source map $f' : X' \longrightarrow Y'$. Either $X' = Q_S(\omega')$, then $Y' = 0$. Or, $X' = Q_S(s)$ for some $s \in S$, then Y' has one-dimensional total space; in particular, Y' is indecomposable. Or, if X' is not of the form $Q_S(s)$ for any $s \in S \cup \{\omega'\}$, then there is an exact sequence (f',g') of S-spaces, say with $g' : Y' \longrightarrow Z'$, the S-space Z' is indecomposable and g' is a sink map.

Of course, as in the case of a module category, the short exact sequences (f,g) with $f : X \longrightarrow Y$ a source map, $g : Y \longrightarrow Z$ a sink map, are called Auslander-Reiten sequences, we write $X = \tau Z$, $Z = \tau^- X$, and there is the same relation between such an Auslander-Reiten sequence and irreducible maps starting in X or ending in Z. In particular, the Auslander-Reiten sequences in $\ell(S)$ show that we may define a translation quiver Γ_S in the same way, as we have defined the Auslander-Reiten quiver of an algebra, and Γ_S is called the Auslander-Reiten quiver of the partially ordered set S. Note that the function $\underline{\dim}_\omega$ is an additive function on Γ_S.

In contrast to the case of an algebra, the Auslander-Reiten quiver of a finite partially ordered set S always has a preprojective component and this component is just ${}_\infty\Gamma_S$. The reason is very simple: the projective vertices of Γ_S are of the form $[P_S(s)]$, $s \in S^+$, there is a unique source in Γ_S, namely $[P_S(\omega)]$, and any other projective vertex has a unique direct predecessor in Γ_S, since, as we have seen, the S-space Y occurring in the sink map $Y \longrightarrow P_S(s)$, for $s \in S$, is indecomposable. The translation quivers of the form ${}_\infty\Gamma_S$ can be characterized as follows.

Given a preprojective translation quiver Γ with a unique source ω, there exists a unique additive function h_Γ such that $h_\Gamma(\omega) = 1$, and $h_\Gamma(p) = \sum_{e(\alpha)=p} h_\Gamma(s(\alpha))$, for any projective vertex $p \neq \omega$ of Γ (the existence and unicity is shown by induction on $d \in \mathbb{N}$, where $\Gamma = {}_d\Gamma$, and then follows for $\Gamma = {}_\infty\Gamma$). A preprojective translation quiver Γ with a unique source is called a <u>left hammock</u> provided h_Γ takes values in $\mathbb{N}_1$ and satisfies $h_\Gamma(q) \geq \sum_{s(\alpha)=q} h_\Gamma(e(\alpha))$, for all injective vertices q of Γ, and, in this case, h_Γ is called the hammock function on Γ. A left hammock Γ is said to be <u>thin</u> provided $h_\Gamma(p) = 1$ for any projective vertex p of Γ.

<u>PROPOSITION</u>. ([RV]) If S is a finite partially ordered set, then ${}_\infty\Gamma_S$ is a thin left hammock, and the hammock function on ${}_\infty\Gamma_S$ is just the function $\underline{\dim}_\omega$. Conversely, any thin left hammock occurs in this way.

In particular, we see that the preprojective component ${}_\infty\Gamma_S$ of Γ_S has no multiple arrows. We may pose the following problem:

<u>Problem 4</u>. Let X,Y be indecomposable S-spaces. Is always $\dim \mathrm{Irr}(X,Y) \leq 1$?

As in the case of an algebra, a component Γ of Γ_S may be said to be <u>standard</u> provided the full additive subcategory of $\ell(S)$ whose indecomposable S-spaces are the S-spaces belonging to Γ is equivalent to $k(\Gamma)$.

<u>PROPOSITION</u>. Let S be a finite partially ordered set. The component ${}_\infty\Gamma_S$ always is standard.

The actual construction of ${}_\infty\ell(S)$ and ${}_\infty\Gamma_S$ is done in the same way as in the case of an algebra. Again, it is convenient to work with integral vectors: given an S-space V, let $\underline{\dim}_S V = (\dim V_\omega; \dim V_s)_{s\in S}$, and we will arrange the entries $\dim V_s$ ($s \in S$) in the form of S. (Recall that the first component $\dim V_\omega$ of $\underline{\dim}_S V$ also has been denoted by $\underline{\dim}_\omega V$.). For example, consider the partially ordered set S

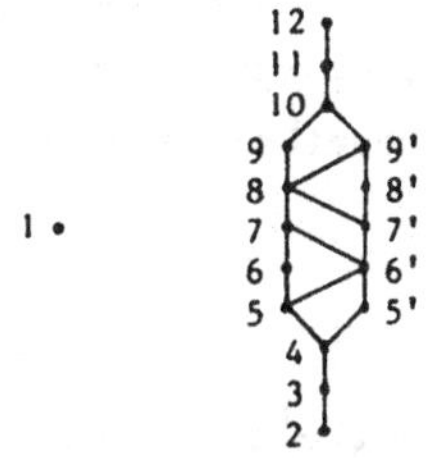

we obtain for ${}_7\Gamma_S$ the following translation quiver:

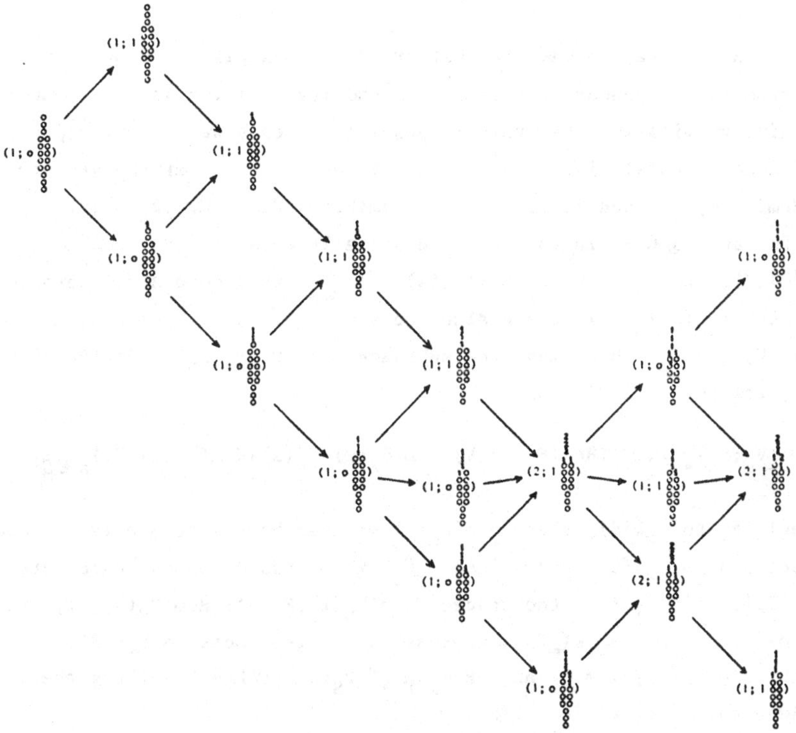

For the example we have chosen here, it turns out that ${}_\infty\Gamma_S$ actually is finite, and (as in the case of an algebra) this implies ${}_\infty\Gamma_S = \Gamma_S$, thus there are only finitely many isomorphism classes of indecomposable S-spaces (of course, in this case, S is said to be representation finite). The complete Auslander-Reiten quiver Γ_S together with its hammock function $\underline{\dim}_\omega$ looks as follows:

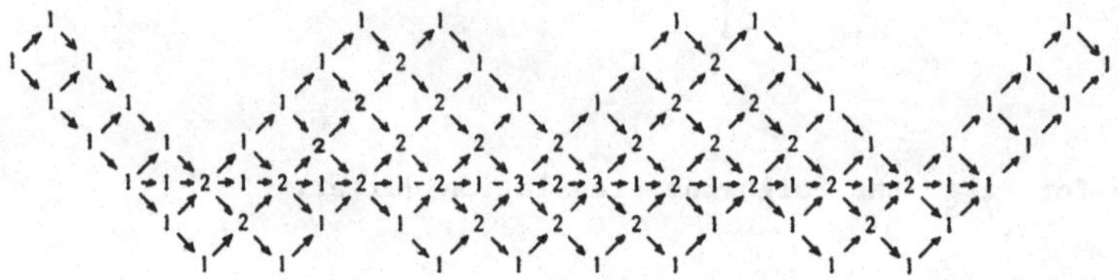

Also here, we may ask whether it is possible to recover an S-space from the component it belongs to, and its position in this component. Again, we will see that this is possible in case we deal with preprojective components! Given any S-space V, we may consider the vectorspace $\mathrm{Hom}(P_S(\omega),V)$ and its subspaces $\mathrm{Hom}(P_S(\omega),P_S(s))\mathrm{Hom}(P_S(s),V)$, defined for any $s \in S$. In this way, we actually obtain an S-space $(\mathrm{Hom}(P_S(\omega),V);\ \mathrm{Hom}(P_S(\omega),P_S(s))\mathrm{Hom}(P_S(s),V))_{s\in S}$. We choose a non-zero element $\xi \in (P_S(\omega))_\omega = k$, the evaluation at ξ gives an isomorphism $\mathrm{Hom}(P_S(\omega),V) \longrightarrow V_\omega$, which maps the subspace $\mathrm{Hom}(P_S(\omega),P_S(s))\mathrm{Hom}(P_S(s),V)$ onto V_s, for any $s \in S$, thus

$$V = (V_\omega;V_s)_{s\in S} \approx (\mathrm{Hom}(P_S(\omega),V);\mathrm{Hom}(P_S(\omega),P_S(s))\mathrm{Hom}(P_S(s),V))_{s\in S}.$$

If V belongs to ${}_\infty\ell(S)$, then $\mathrm{Hom}(P_S(s),V)$ can be non-zero only in case $P_S(s)$ belongs to ${}_\infty\ell(S)$, and, since ${}_\infty\Gamma_S$ is standard, we can calculate for any $P_S(s)$ in ${}_\infty\ell(S)$ the spaces $\mathrm{Hom}(P_S(\omega),P_S(s)),\mathrm{Hom}(P_S(s),V)$, and the composition, inside $k({}_\infty\Gamma_S)$, whereas for $P_S(s)$ outside ${}_\infty\ell(S)$, we have both $\mathrm{Hom}(P_S(s),V) = 0$ and $\mathrm{Hom}_{k(\Gamma_S)}([P_S(s)],[V]) = 0$. Altogether, we conclude that for V in ${}_\infty\Gamma(S)$, we have

$$V = (V_\omega;V_s)_{s\in S} \approx (\mathrm{Hom}_{k(\Gamma_S)}(p(\omega),v);\mathrm{Hom}_{k(\Gamma_S)}(p(\omega),p(s))\mathrm{Hom}_{k(\Gamma_S)}(p(s),v))_{s\in S}$$

where $p(s) = [P_S(s)]$ for any $s \in S^+$, and $v = [V]$. Of course, we should keep in mind that the partially ordered set S itself can be recovered from $k(\Gamma)$ only partially, what we can recover is the full subset $\{s \in S \mid P_S(s) \in {}_{\infty}\ell(S)\}$, it is a filter in S. Given any S-space V, let us define its <u>support</u> S_V as the set of all $s \in S$ such that $\sum_{t<s} V_t \subsetneqq V_s$, with the induced ordering. Clearly, we may recover V if we know just V_ω and the subspaces V_s, $s \in S_V$. For example, in the special case considered above, the S-space $V = \tau^{-6}P_S(10)$ has as support the subset

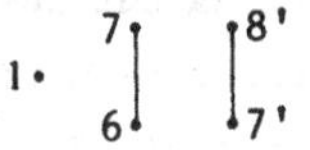

of S, so it is completely determined by its total space which we may assume to be $\mathrm{Hom}_{k(\Gamma_S)}(p(\omega),v)$, and the subspaces $\mathrm{Hom}_{k(\Gamma_S)}(p(\omega),p(s))\mathrm{Hom}_{k(\Gamma_S)}(p(s),v)$, for $s = 1, 6, 7, 7'$, and $8'$. Let us display the vertices $p(s)$ for $s = \omega, 1, 6, 7, 6', 8'$; and, in addition, the vertex v inside Γ_S:

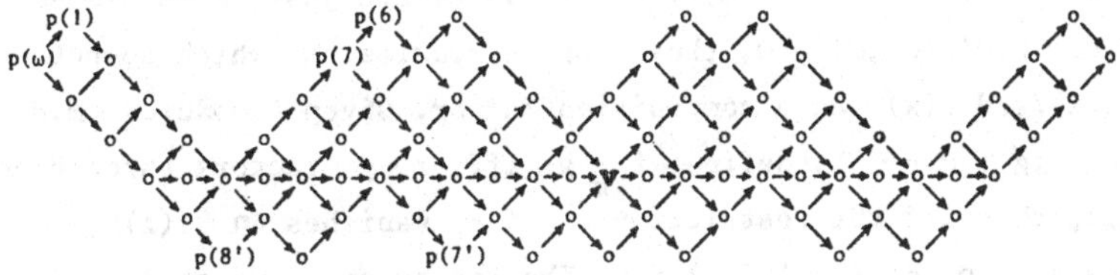

Note that the dimension vector of V, considered as an S_V-space, is $(3;2\,{}^{22}_{11})$.

A finite left hammock is just called a <u>hammock</u>. For left hammocks, there is the following result on the growth of the hammock function along a τ-orbit, and we will outline below that this implies that any hammock is thin.

<u>PROPOSITION</u>. ([RV]) Let H be a left hammock, and x a vertex of H. If x is injective, then $h_H(x) = 1$. If x is not injective, then $h_H(x) \leq h_H(\tau^- x) + 1$.

This result has many consequences. First of all, given an injective vertex q of a left hammock H, then q can be starting point of at most one arrow, and if there is an arrow $q \longrightarrow y$, then also $h_H(y) = 1$. Consequently, a hammock has a unique sink. It follows easily that the opposite translation quiver H^* of a hammock H, is a left hammock again, and in fact, a thin one. Thus, by symmetry, a hammock is always thin. We conclude that the hammocks are just the Auslander-Reiten quivers of the representation-finite partially ordered sets.

As an application, let us outline a relation between the Auslander-Reiten quivers of representation-finite algebras and representation-finite partially ordered sets. In fact, first consider only representation-directed algebras; by definition, these are those representation-finite algebras A whose Auslander-Reiten quiver Γ_A has no oriented cycles, or equivalently, those algebras A for which Γ_A is a preprojective translation quiver. Let P(x) be an indecomposable projective A-module, say P(x) = Ae(x) for some primitive idempotent e(x) in A, and let $A_x = A/\langle e(x)\rangle$, where $\langle e(x)\rangle$ is the twosided ideal of A generated by e(x). Of course, the A_x-modules are just those A-modules M satisfying Hom(P(x),M) = 0, thus those A-modules M which do not contain E(x) = P(x)/rad P(x) as a composition factor. Given A-modules M,N and two maps $f,g : M \longrightarrow N$, define $f \underset{x}{\sim} g$ iff $f-g$ factors through an A_x-module, thus iff the restriction of $f-g$ vanishes on e(x)M, and this is equivalent to Hom(P(x),f-g) = 0. The factor category of A-mod obtained in this way is denoted by $\mathcal{H}(x)$, its objects are the same as those of A-mod, and

$$\mathrm{Hom}_{\mathcal{H}(x)}(M,N) = \mathrm{Hom}(M,N)/\underset{x}{\sim}\ .$$

Note that the indecomposable objects in $\mathcal{H}(x)$ are given by those indecomposable modules M with Hom(P(x),M) ≠ 0. If we denote by H(x) the full translation subquiver of Γ_A with vertices of the form [M], where M is an indecomposable module satisfying Hom(P(x),M) ≠ 0, then one sees rather easily that $\mathcal{H}(x)$ is equivalent to k(H(x)), since A-mod is equivalent to $k(\Gamma_A)$. Let us investigate the translation quiver H(x): With Γ_A also H(x) is a preprojective translation quiver, and obviously H(x) has a unique source, namely [P(x)]. Consider the function $[M] \longrightarrow \dim \mathrm{Hom}(P(x),M)$ on H(x). Since P(x) is projective, this is an additive function even on all of Γ_A. Also dim Hom(P(x),P,(x)) = 1, whereas for P(y) indecomposable projective, and $[P(y)] \neq [P(x)]$, we have $\dim \mathrm{Hom}(P(x),P(y)) = \dim \mathrm{Hom}(P(x), \mathrm{rad}\, P(y))$. It follows that this function is just $h_{H(x)}$. Consequently, $h_{H(x)}$ takes values in $\mathbb{N}_1$. Also, for Q indecomposable injective, we have $\dim \mathrm{Hom}(P(x),Q) \geq \dim \mathrm{Hom}(P(x),Q/\mathrm{soc}\, Q)$, thus $h_{H(x)}$ satisfies all the properties of a hammock function, thus $H(x)$ is a finite left hammock, thus a hammock. We can now use the previous result which asserts that $H(x)$ is the Auslander-Reiten quiver $\Gamma_{S(x)}$ of some representation finite partially ordered set. Thus, there is the following theorem:

__THEOREM__. ([RV]) For any indecomposable projective module $P(x)$ of a representation-directed algebra, there exists a partially ordered set $S(x)$ such that $\mathcal{H}(x)$ and $\ell(S(x))$ are equivalent categories, and $H(x)$ and $\Gamma_{S(x)}$ isomorphic translation quivers.

Note that the hammock function on $H(x)$ is $[M] \longmapsto \dim \mathrm{Hom}(P(x),M)$, whereas the hammock function on $\Gamma_{S(x)}$ is $[V] \longmapsto \underline{\dim}_\omega V$, and the unicity of hammock functions asserts that these functions are equal when we identify $H(x)$ and $\Gamma_{S(x)}$. There is the following corollary:

__COROLLARY__. If M is an indecomposable module over a representation-directed algebra, then all entries of the dimension vector $\underline{\dim}\, M$ are bounded by 6.

__Proof__. The entries of $\underline{\dim}\, M$ are $(\underline{\dim}\, M)_x = \dim \mathrm{Hom}(P(x),M)$; thus, if $(\underline{\dim}\, M)_x \neq 0$, then $[M]$ is in $H(x)$, and $(\underline{\dim}\, M)_x = h_{H(x)}([M])$. On the other hand, a well-known theorem of Klejner asserts that the total space of an indecomposable S-space, where S is representation finite, is bounded by 6. Since $S(x)$ is representation finite, we can use this theorem: the hammock function on $\Gamma_{S(x)}$ is bounded by 6.

The rather strange bound 6 first appeared in the theorem of Klejner. There is a general result due to Ovsienko from which one may deduce the assertion of the corollary without difficulty. There are also other proofs of the corollary known which however seem to be more awkward. Note that the proof above shows that the assertion concerning modules is a direct consequence of that concerning S-spaces.

Assume now that A is basic, let $_A A = \bigoplus_{x=1}^{n} P(x)$, thus $P(1),\ldots,P(n)$ is a complete list of the indecomposable projective modules. For any x, let $S(x)$ be the partially ordered set given by the theorem, with $\mathcal{H}(x) \approx \ell(S(x))$, and $H(x) \approx \Gamma_{S(x)}$. If M is an indecomposable module, let $M_x = \mathrm{Hom}(P(x),M)$, and we identify M with $\bigoplus M_x$. If M is indecomposable and $M_x \neq 0$, then M is an indecomposable object in $\mathcal{H}(x) \approx \ell(S(x))$, and the dimension of the $S(x)$-space $\bar{M}_x$ corresponding to M is $\dim M_x$, thus we may consider M_x itself as total space of this $S(x)$-space, thus $\bar{M}_x = (M_x;(\bar{M}_x)_s)_{s \in S(a)}$. Note the following:

1) We obtain all indecomposable $S(x)$-spaces in the form $\bar{M}_x$, with M indecomposable (and $M_x \neq 0$).
2) If M is indecomposable, $M_x \neq 0$, then the isomorphism class of the $S(x)$-space $\bar{M}_x$ determines the isomorphism class of M uniquely.
3) Let $n(x)$ be the sum of the dimensions of the total spaces of all $S(x)$-spaces occurring in a complete list of indecomposable $S(x)$-spaces (thus, the sum of all values of the hammock function $h_{H(x)}$). Then $\sum_{x=1}^{n} n(x)$ is the sum of the dimensions of the modules occurring in a complete list of indecomposable modules.

Indeed the direct sum $\bigoplus M$ of all modules M occurring in a complete list of indecomposable modules can be written in the form $\bigoplus_M \bigoplus_x M_x$, and the dimension of $\bigoplus_M M_x$ ist just $n(x)$.

As an illustration of the theorem, we consider again the algebra A which has served before as example; as we have seen, this algebra is representation-directed. If we focus on the hammock $H(b)$ defined by the vertex b, we obtain the following shaded part of Γ_A:

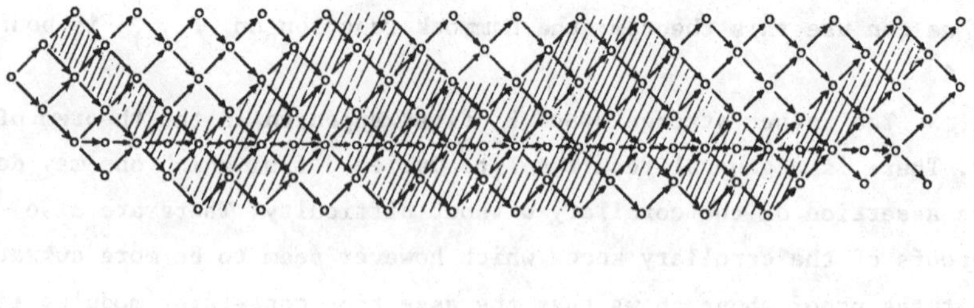

but this is just the hammock which was exhibited above as the Auslander-Reiten quiver Γ_S of some specific partially ordered set, thus $S = S(b)$.

In which way does one obtain $S(x)$ in general? The elements of $S(x)$ are the projective vertices of $H(x)$, thus they are of the form $[U]$, where U is an indecomposable module, with $\mathrm{Hom}(P(x),U) \neq 0$, and $\mathrm{Hom}(P(x),\tau U) = 0$. The fact that the hammock $H(x)$ is thin means just that for such a module U, the space $\mathrm{Hom}(P(x),U)$ is 1-dimensional. Given two indecomposable modules U,U' with $[U],[U']$ in $S(x)$, we have $[U] \geq [U']$ in $S(x)$ if and only if $\mathrm{Hom}_{H(x)}(U,U') \neq 0$, thus if and only if $\mathrm{Hom}(P(x),U)\mathrm{Hom}(U,U') \neq 0$.

Similarly, given an indecomposable module M with $M_x = \mathrm{Hom}(P(x),M) \neq 0$, and $[U] \in S(x)$, we can write down $(\bar{M}_x)_{[U]}$ as follows:

$$(\bar{M}_x)_{[U]} = \mathrm{Hom}(P(x),U)\mathrm{Hom}(U,M) \subseteq \mathrm{Hom}(P(x),M) = M_x.$$

Take a projective presentation of U,

$$\bigoplus_{i=1}^{r} P(y_i) \xrightarrow{[\gamma_{ij}]_{ij}} \bigoplus_{j=o}^{t} P(x_j) \xrightarrow{[\varepsilon_j]_j} U \longrightarrow 0$$

with $x_0 = x$, and $\varepsilon_0 \neq 0$ (note that, in general, a minimal projective presentation will not satisfy these conditions!). Then

$$(\bar{M})_{[U]} = \{m_0 \in M_x \mid \exists\, m_j \in M_{x_j},\ 1 \leq j \leq t \text{ with } \sum_{j=0}^{t} \gamma_{ij}\, m_j = 0, \text{ for all } i\}.$$

Let us outline the proof: since $\mathrm{Hom}(P(x),U)$ is 1-dimensional, $\mathrm{Hom}(P(x),U) = k\varepsilon_0$, and therefore $\mathrm{Hom}(P(x),U)\mathrm{Hom}(U,M) = \varepsilon_0 \mathrm{Hom}(U,M)$. Our presentation induces an exact sequence

$$0 \longrightarrow \mathrm{Hom}(U,M) \longrightarrow \bigoplus_{j=o}^{t} \mathrm{Hom}(P(x_j),M) \longrightarrow \bigoplus_{i=1}^{r} \mathrm{Hom}(P(y_i),M).$$

Here, $\eta \in \mathrm{Hom}(U,M)$ is mapped to the tuple $(\varepsilon_0\eta,\ldots,\varepsilon_t\eta)$ which satisfies $\sum_j \gamma_{ij}\, \varepsilon_j\eta = 0$ for all i. Now, any element in $\mathrm{Hom}(P(x),U)\mathrm{Hom}(U,M)$ can be written in the form $\varepsilon_0\eta$ with $\eta \in \mathrm{Hom}(U,M)$, thus define $m_j = \varepsilon_j\eta$. Conversely, assume there are given $m_j' \in M_{x_j}$ with $\sum_{j=0}^{t} \gamma_{ij}\, m_j' = 0$, for all i. The exactness of the sequence above gives $\eta' \in \mathrm{Hom}(U,M)$ with

$m_j' = \varepsilon_j \eta'$ in particular, m_o' belongs to $\varepsilon_o \mathrm{Hom}(U,M) = \mathrm{Hom}(P(x),U)\mathrm{Hom}(U,M)$.

We return to our specific example, and consider the A-module $M = \tau^{-7}P(e)$, with dimension vector $\underline{\dim}\, M = \begin{smallmatrix} & & 2 & & \\ & & 3 & & \\ 2 & 3 & 4 & 3 & 2 \\ & & 2 & & \end{smallmatrix}$. We want to determine the $S(b)$-space $\bar{M}_b$. A comparison of the position of $[M]$ in the hammock $H(b)$ reveals that $[M]$, considered as an element of $H(b) = \Gamma_{S(b)}$ is just $[\tau^{-6}P_{S(b)}(10)]$. For any $s \in S(b)$, let $U(s)$ be the indecomposable module with $[U(s)]$ in $H(b)$, and $[U(s)] = [P_S(s)]$ as an element of $H(b) = \Gamma_{S(b)}$. We recall that the support of $\tau^{-6}P_{S(b)}(10)$ is the full subset of $S(b)$ given by the elements $\{1,6,7,7',8'\}$, and that the dimension vector of $\tau^{-6}P_{S(b)}(10)$, restricted to its support, is $(3;2\begin{smallmatrix}22\\11\end{smallmatrix})$. We are dealing with the following displayed modules:

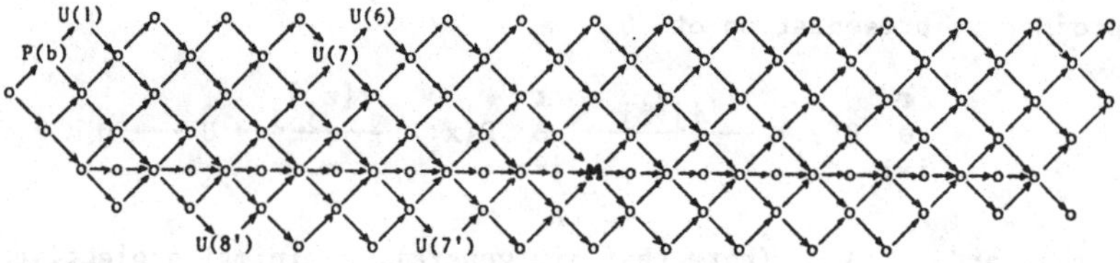

The recipe for determining $(\bar{M}_b)_{[U(s)]}$ easily yields the following:

s	$\underline{\dim}\, U(s)$	$(\bar{M}_b)_{[U(s)]}$	s	$\underline{\dim}\, U(s)$	$(\bar{M}_b)_{[U(s)]}$
1	$\begin{smallmatrix} & & 1 & & \\ & & 1 & & \\ 1 & 0 & 0 & 0 & 0 \\ & & 0 & & \end{smallmatrix}$	γM_c	8'	$\begin{smallmatrix} & & 0 & & \\ & & 1 & & \\ 1 & 1 & 1 & 1 & 1 \\ & & 0 & & \end{smallmatrix}$	$\alpha_2 M_h$
7	$\begin{smallmatrix} & & 0 & & \\ & & 1 & & \\ 1 & 1 & 1 & 1 & 0 \\ & & 1 & & \end{smallmatrix}$	$\alpha^{-1}\beta_1\beta_2(\beta_3 M_f \cap \delta M_g)$	7'	$\begin{smallmatrix} & & 0 & & \\ & & 1 & & \\ 1 & 1 & 2 & 1 & 1 \\ & & 1 & & \end{smallmatrix}$	$\alpha_2\beta_4^{-1}\beta_3^{-1}(\beta_2^{-1}O_d + M_g)$
6	$\begin{smallmatrix} & & 0 & & \\ & & 1 & & \\ 0 & 0 & 0 & 0 & 0 \\ & & 0 & & \end{smallmatrix}$	$\alpha_1^{-1}O_a$			

For example, in dealing with U(7), we use the presentation

$$P(a) \oplus P(e) \xrightarrow{\begin{bmatrix}\alpha_1 & -\beta_1\beta_2\beta_3 & 0\\ 0 & \beta_3 & -\delta\end{bmatrix}} P(b) \oplus P(f) \oplus P(g) \longrightarrow U(7) \longrightarrow 0,$$

thus $(\bar{M}_b)_{[U(7)]}$ is given by the elements $x_0 \in M_b$ such that there exist $x_1 \in M_f$, $x_2 \in M_g$ satisfying

$$\alpha_1 x_0 - \beta_1\beta_2\beta_3 x_1 = 0, \quad \beta_3 x_1 - \delta x_2 = 0.$$

It follows that $x = x_0 \in M_b$ belongs to $(\bar{M}_b)_{[U(7)]}$ if and only if

$$x \in \alpha_1^{-1}\beta_1\beta_2\beta_3\beta_3^{-1}\delta(M_g) = \alpha_1^{-1}\beta_1\beta_2(\beta_3 M_f \cap \delta M_g).$$

Altogether, we see that the hammock approach to representation directed algebras leads to an intrinsic interpretation of the use of subspace methods.

Let us add that the hammock approach is not restricted to representation-directed algebras. In case we deal with a representation-finite algebra A, we may use the well-established covering techniques. Actually, we will consider an arbitrary algebra A and an indecomposable projective A-module P(a) provided there are only finitely many isomorphism classes of indecomposable modules M with $\mathrm{Hom}(P(a),M) \neq 0$. Of course, in this case, all indecomposable modules M with $\mathrm{Hom}(P(a),M) \neq 0$ belong to a single component. We use the filtration of Hom(P(a),M) given by the subspaces $M(d) = \mathrm{rad}^d(P(a),M)$. The dimension of $M(d)/M(d+1)$ counts the multiplicity of Jordan-Hölder factors of M of the form P(a)/rad P(a) which can be reached from P(a) by means of maps in rad^d, but not in rad^{d+1}. We define a translation quiver H(a) as follows: its vertices are the pairs ([M],d), where M is an indecomposable module with $M(d) \neq M(d+1)$. There are only arrows $([M],d) \longrightarrow ([N],e)$ for $e = d+1$, and the number of arrows is equal to $\dim_k \mathrm{Irr}(M,N)$. Finally, ([M],d) is projective if either $d \le 1$, or both $d \ge 2$ and $(\tau M)(d-2) \neq (\tau M)(d-1)$; and $\tau([M],d) = ([\tau M],d-2)$, otherwise. For A representation-directed, the translation quiver constructed in this way is canonically isomorphic to the previously constructed hammock. Always, the translation quiver H(a) is a hammock, thus, there exists a representation-finite partially ordered set S(a) with $H(a) = \Gamma_{S(a)}$.

We end this lecture with two problems.

Problem 5. Which representation-finite partially ordered sets do occur in the form S(a)?

Note that there are examples of representation-finite partially ordered sets which cannot be realized in the form S(a).

Problem 6. Let A be a representation-finite algebra. Is it possible to construct the partially ordered sets S(a) without prior knowledge of Γ_A? And, what kind of relations do exist between the various S(a)?

LECTURE 2

SEPARATING SUBCATEGORIES: CONNECTING COMPONENTS AND SEPARATING TUBULAR FAMILIES

A full subcategory of A-mod which is closed under direct sums and direct summands will be called a <u>module class.</u> We will focus our attention to an interesting phenomenon, the existence of separating subcategories: for some kinds of algebras A, it turns out that there exist module classes $\mathcal{P}$, $\mathcal{S}$, $\mathcal{Q}$ in A-mod such that any indecomposable A-module belongs to $\mathcal{P}$ or $\mathcal{S}$ or $\mathcal{Q}$ (thus, $\mathcal{P}$, $\mathcal{S}$, $\mathcal{Q}$ exhaust the category A-mod), that $\mathrm{Hom}(X,Y) = 0$ provided one of the following conditions is satisfied: $X \in \mathcal{Q}$, $Y \in \mathcal{P}$; or $X \in \mathcal{Q}$, $Y \in \mathcal{S}$; or $X \in \mathcal{S}$, $Y \in \mathcal{P}$, and finally, given $P \in \mathcal{P}$, $Q \in \mathcal{Q}$, then any map $P \longrightarrow Q$ can be factored through an object in $\mathcal{S}$. In this case, we call $\mathcal{S}$ a <u>separating</u> subcategory, separating $\mathcal{P}$ from $\mathcal{Q}$. We may visualize these subcategories $\mathcal{P}$, $\mathcal{S}$, $\mathcal{Q}$ in the following way

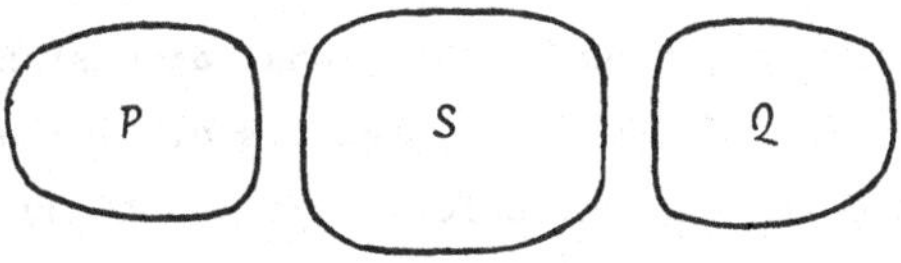

with possible maps only going from left to right.

We are going to list some properties of separating subcategories. In order to do so, we will need some further definitions. A sequence $(X_0,X_1,\ldots,X_m)$ of indecomposable modules with $\mathrm{rad}(X_{i-1},X_i) \neq 0$, for all $1 \leq i \leq m$, will be called a <u>path</u> of length m in A-mod. In case there exists a path $(X_0,X_1,\ldots,X_m)$ in A-mod, we will write $X_0 \preceq X_m$. Note that for any path $[X_0] \longrightarrow [X_1] \longrightarrow \ldots \longrightarrow [X_m]$ in Γ_A, the sequence $(X_0,X_1,\ldots,X_m)$ is a path in A-mod, whereas, obviously, the converse usually will not be true. Note that for a given path $(X_0,X_1,\ldots,X_m)$, the modules X_i may not even belong to the same component of Γ_A. A subcategory $\mathcal{M}$ of A-mod is said to be <u>path-closed</u> provided for any path $(X_0,X_1,\ldots,X_m)$ in A-mod, with X_0,X_m in $\mathcal{M}$, all X_i belong to $\mathcal{M}$. A module M is said to be <u>sincere</u> provided any simple module occurs as a composition factor of M. Note that any faithful module is sincere, but it is easy to construct sincere modules which are not faithful (unless A is semisimple): just take the direct sum of all simple modules. In general, there even will exist indecomposable sincere modules which are not faithful.

Assume now that $\mathcal{S}$ separates $\mathcal{P}$ from $\mathcal{Q}$. Then $\mathcal{P}$, $\mathcal{S}$, $\mathcal{Q}$ all are path closed, P is closed under τ, whereas Q is closed under τ^-. Also, S is closed under τ if and only if P is closed under τ^-, and S is closed under τ^- if and only if Q is closed under τ. If A is connected, and $S \neq \{0\}$, then P and Q are uniquely determined by S. Assume now in addition that all projective modules belong to P and all injective ones to Q. Then S contains sincere modules, and rather strong homological properties are satisfied by P, S, Q. Namely, if X belongs to P or to S, then proj.dim.$X \leq 1$. (The proof rests on the well-known criterion that proj.dim $M \leq 1$ if and only if $\mathrm{Hom}(I,\tau M) = 0$ for any indecomposable injective module I). Similarly, if Y belongs to S or to Q, then inj.dim. $Y \leq 1$. Since any submodule of a projective module belongs to P, and the modules in P have projective dimension at most 1, we see that proj.dim $M \leq 2$ for any module M, thus the global dimension of A is at most 2.

The algebras which we will exhibit in this lecture often will have finite global dimension, usually the global dimension will be rather small. One of the important features of algebras of finite global dimension is the possibility of using quadratic forms. We denote by $K_0(A)$ the Grothendieck group of all A-modules modulo all short exact sequences. It is a free abelian group of finite rank, with the set of simple modules as a basis. With respect to this basis, we may identify $K_0(A)$ with $\mathbb{Z}^n$, and the element of $K_0(A)$ corresponding to the module M is just $\underline{\dim}\, M$. Assume now that A has finite global dimension, say gl.dim.$A = d$. Then it is easy to see that

$$\langle \underline{\dim}\, M, \underline{\dim}\, N\rangle = \sum_{i=0}^{d} (-1)^i \dim \mathrm{Ext}^i(M,N)$$

(with $\mathrm{Ext}^0 = \mathrm{Hom}$) is well defined and extends to a bilinear form on $K_0(A)$. The corresponding quadratic form will be denoted by $\chi = \chi_A$, thus $\chi(x) = \langle x,x\rangle$ for $x \in K_0(A)$, but one should observe that $\langle -,-\rangle$ usually is non-symmetric. There is an easy way to calculate $\langle -,-\rangle$. Denote by P_i the projective cover of the simple module S_i, $1 \leq i \leq n$, and let C_A be the $n \times n$-matrix with i-j-entry given by $\dim \mathrm{Hom}(P(i),P(j))$, thus the j-th column is just $(\underline{\dim}\, P(j))^T$. This matrix C_A is called the Cartan matrix for A. For A of finite global dimension, C_A is invertible (even over $\mathbb{Z}$), and

$$\langle x,y\rangle = x C^{-T} y^T$$

for $x,y \in K_0(A)$.

The first examples of non-trivial separating subcategories to be known were found for hereditary algebras. Let us dwell for a moment on these algebras, so suppose A is a (finite-dimensional) connected hereditary algebra, thus the path algebra $A = k\Delta$ of some finite connected quiver without oriented cycles. As we have mentioned in the first lecture, A always has a preprojective component and this component is unique and embeds into $\mathbb{Z}\Delta$. We denote by $\mathcal{P}$ the module class whose indecomposable modules belong to the preprojective component; the modules in $\mathcal{P}$ will be said to be preprojective. Now, $k\Delta$ is representation-finite if and only if Δ is of type $\mathbb{A}_n$, $\mathbb{D}_n$, $\mathbb{E}_6$, $\mathbb{E}_7$, or $\mathbb{E}_8$, and in this case, $\mathcal{P} = A\text{-mod}$. So assume $k\Delta$ is not representation-finite. Then, the preprojective component is of the form $\mathbb{N}\Delta$ (= the full translation subquiver of $\mathbb{Z}\Delta$ given by all vertices (z,x) with $z \in \mathbb{N}$). By duality, there is a similarly defined module class $\mathcal{Q}$, modules in $\mathcal{Q}$ will be said to be preinjective (with $k\Delta$, also $k\Delta^*$ is a hereditary algebra, and a $k\Delta$-module belongs to $\mathcal{Q}$ if and only if its k-dual is a preprojective $k\Delta^*$-module), and the component of Γ_A containing the indecomposable preinjective modules is of the form $(-\mathbb{N})\Delta$. There always are additional modules which have no indecomposable summand in either $\mathcal{P}$ or $\mathcal{Q}$, they will be said to be regular, and we denote by $\mathcal{R}$ the full subcategory of all regular modules. Then $\mathcal{R}$ is a separating subcategory, separating $\mathcal{P}$ from $\mathcal{Q}$. Taking into account the particular shape of the preprojective and the preinjective component, we may think of A-mod as being of the following form.

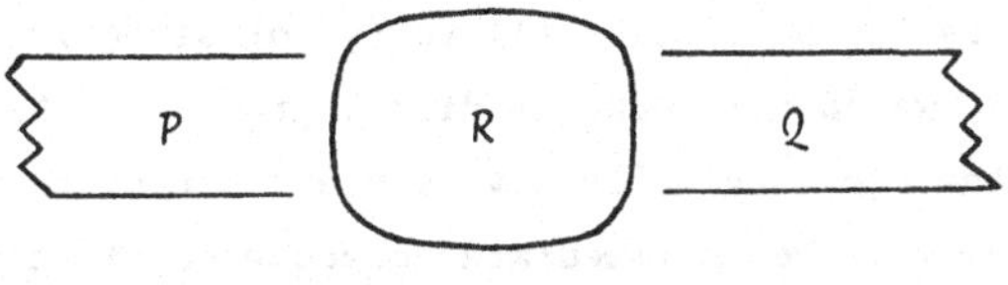

Of course, the regular components contain only regular modules, and the indecomposable regular modules all belong to regular components. We have mentioned in the first lecture that for these regular components, there are only very restrictive possibilities: they are regular tubes, in case Δ is an Euclidean quiver ($\tilde{\mathbb{A}}_n$, $\tilde{\mathbb{D}}_n$, $\tilde{\mathbb{E}}_6$, $\tilde{\mathbb{E}}_7$ or $\tilde{\mathbb{E}}_8$), or else they are of the form $\mathbb{Z}\mathbb{A}_\infty$. Note that $k\Delta$ is tame if and only if Δ is Euclidean. The actual construction of the regular tubes in this case will be outlined later in this lecture.

Consider now again an arbitrary finite-dimensional algebra A. An indecomposable module M will be said to be directing provided there does not exist a path $(X_0, X_1, \ldots, X_n)$ of length $n \geq 1$ with $X_0 = M = X_n$. Directing modules have very pleasant properties: Let M be directing, then:

(1) $\operatorname{End}(M) = k$, $\operatorname{Ext}^i(M,M) = 0$ for all $i \geq 1$.

A direct consequence of (1) is the following:

(1') If A has finite global dimension (so that χ_A is defined), $\chi_A(\underline{\dim}\, M) = 1$.

(2) $\underline{\dim}\, M$ determines M uniquely (more precisely: if also N is indecomposable, and $\underline{\dim}\, M = \underline{\dim}\, N$, then M and N are isomorphic; note that the module N is not assumed to be directing).

(3) The annihilator of M in A is generated, as an ideal, by idempotents. As a consequence, if M is sincere, then M is actually faithful.

(4) If M is sincere, then $\text{proj.dim.} M \leq 1$, $\text{inj.dim.} M \leq 1$, and $\text{gl.dim.} A \leq 2$.

(5) If M is sincere, there are only finitely many indecomposable modules X with both $X \not\preceq M$ and $M \not\preceq X$.

Of course, one should keep in mind that for any indecomposable module N, we may look at the classes $\mathcal{P}(N) = \{X \mid X \preceq N\}$ of all predecessors and $\mathcal{Q}(N) = \{X \mid N \preceq X\}$ of all successors of N with respect to $\preceq$, and N will be directing if and only if N is the only module which belongs both to $\mathcal{P}(N)$ and to $\mathcal{Q}(N)$. The property (5) now asserts that for M sincere and directing, all but at most finitely many indecomposable modules will belong to $\mathcal{P}(N) \cup \mathcal{Q}(N)$.

Whereas the properties (1) to (4) of directing modules can be verified directly, we do not know any direct proof for (5). We will investigate below the algebras which have sincere directing modules rather carefully, and (5) will be an immediate consequence of these investigations; actually we will obtain an explicit bound. The main technical tool will be the tilting theory.

What are examples of directing modules? First of all, any indecomposable module in ${}_{\infty}(A\text{-mod})$ is directing: in particular, any indecomposable module belonging to a preprojective component is directing. Of course, there is the dual assertion: any indecomposable module belonging to a "preinjective" component is directing. We later will see that it is possible to construct also regular components which only contain directing modules. First, let us explain the main principles of tilting theory.

By definition, a <u>tilting module</u> is a module T satisfying the following three properties: its projective dimension is at most 1, the T-codimension of ${}_AA$ is at most one, and $\mathrm{Ext}^1(T,T) = 0$ (we say that T-codim $M \leq m$ provided there exists an exact sequence

$$0 \longrightarrow M \longrightarrow M_0 \longrightarrow M_1 \longrightarrow \dots \longrightarrow M_m \longrightarrow 0$$

with all M_i being direct sums of direct summands of T); in the definition of a tilting module the second condition may be replaced by the following: T has at least n pairwise non-isomorphic indecomposable direct summands where n is the number of simple A-modules (actually, the other conditions force that such a module can have at most n pairwise non-isomorphic indecomposable direct summands!) We should indicate that these conditions refer to T as an A-module, and write ${}_AT$ instead of T, since we will have to deal at the same time with T considered as a right B-module T_B, where $B = \mathrm{End}({}_AT)$. There is also the dual notion of a cotilting module S (its injective dimension is at most one, the S-dimension of any injective cogenerator is at most one, and $\mathrm{Ext}^1(S,S) = 0$).

Given any module M, let $G(M)$ be the full subcategory of all modules generated by M, and $C(M)$ that of all modules cogenerated by M. Also recall that a pair (F,T) of full subcategories of A-mod is said to be a torsion pair provided F consists of all modules X with $\mathrm{Hom}(Y,X)=0$ for all $Y \in T$, and T consists of all modules Y with $\mathrm{Hom}(Y,X) = 0$ for all $X \in F$; in this case, the modules in F are said to be torsion free, those in T are said to be torsion (observe that our notation (F,T) of a torsion pair presents first the class of torsion free, then the class of torsion modules, in contrast to a fairly standard usage in ring theory, however our notation seems to reflect better the underlying theme of "nonzero maps going from left to right"). The torsion pair (F,T) is said to be split provided any indecomposable module belongs to F or to T. Now we may formulate the main result of tilting theory:

THEOREM (Brenner, Butler). Let ${}_AT$ be a tilting module, and $B = \mathrm{End}({}_AT)$. Then the k-dual ${}_BS = D(T_B)$ of T_B is a cotilting module, $\mathrm{End}(T_B) = A$, the pairs $(C(\tau_A T), G({}_AT))$ and $(C({}_BS), G(\tau_B^- S))$ are torsion pairs, and $G({}_AT)$ is equivalent to $C({}_BS)$, and $C(\tau_A T)$ is equivalent to $G(\tau_B^- S)$.

Let us write down explicitly functors which provide the equivalences asserted in the theorem, note that they are restrictions of functors which are defined on the complete module categories. First of all, there is the functor $\mathrm{Hom}_A({}_AT_B,-)$: A-mod $\longrightarrow$ B-mod, it vanishes on $\mathcal{C}(\tau_AT)$, has image $\mathcal{C}({}_BS)$, and its restriction to $\mathcal{G}({}_AT)$ gives the equivalence $\mathcal{G}({}_AT) \longrightarrow \mathcal{C}({}_BS)$. Similarly, there is the functor $\mathrm{Ext}^1_A({}_AT_B,-)$: A-mod $\longrightarrow$ B-mod, it vanishes on $\mathcal{G}({}_AT)$, has image $\mathcal{G}(\tau_B^-S)$, and its restriction to $\mathcal{C}(\tau_AT)$ gives the second equivalence $\mathcal{C}(\tau_AT) \longrightarrow \mathcal{G}(\tau_B^-S)$. As reverse functors, use the restriction of ${}_AT_B\otimes-$ to $\mathcal{C}({}_BS)$, and of $\mathrm{Tor}^B_1({}_AT_B,-)$ to $\mathcal{G}(\tau_B^-S)$.

<u>PROPOSITION:</u> Let A be a hereditary algebra, ${}_AT$ a tilting module, $B = \mathrm{End}({}_AT)$, and ${}_BS = D(T_B)$. Then the torsion pair $(\mathcal{C}({}_BS), \mathcal{G}(\tau_B^-S))$ is split.

An algebra of the form $B = \mathrm{End}({}_AT)$, with ${}_AT$ a tilting module, and A hereditary, will be called a <u>tilted</u> algebra. The proposition above asserts that in this situation, we can recover all indecomposable B-modules N from indecomposable A-modules. For, if N belongs to $\mathcal{C}({}_BS)$, then $N = \mathrm{Hom}_A({}_AT_B, X)$ for some indecomposable A-module X in $\mathcal{G}({}_AT)$, whereas, if N does not belong to $\mathcal{C}({}_BS)$, then N belongs to $\mathcal{G}(\tau_B^-S)$, and therefore $N = \mathrm{Ext}^1_A({}_AT_B, X)$ for some indecomposable A-module X in $\mathcal{C}(\tau_AT)$. In this way, we obtain a bijection between the indecomposable B-modules and certain indecomposable A-modules, namely those which are either torsion, or torsion free with respect to our torsion pair.

There is the following criterion for an algebra B in order to be a tilted algebra. A module class $\mathcal{S}$ in B-mod will be called a <u>slice</u> provided $\mathcal{S}$ is path closed, contains a sincere module, and if, in addition, the following property is satisfied: given any Auslander-Reiten sequence $0 \to X \to Y \to Z \to 0$ in B-mod, then at most one of X,Z belongs to $\mathcal{S}$, and one of X,Z belongs to $\mathcal{S}$ in case an indecomposable direct summand of Y is in $\mathcal{S}$.

The module category of a tilted algebra always contains a slice: Let A be hereditary, ${}_AT$ a tilting module, and $B = \mathrm{End}({}_AT)$. Denote by $\mathcal{S}$ the set of B-modules of the form $\mathrm{Hom}_A({}_AT_B, I)$, where I is an injective A-module. Then $\mathcal{S}$ is a slice in B-mod. We may describe $\mathcal{S}$ also alternatively: Let ${}_BS = D(T_B)$. Then $\mathcal{S}$ is the set $\langle {}_BS \rangle$ of all modules which are direct sums of direct summands of ${}_BS$, therefore ${}_BS$ is called a <u>slice module</u>. (Thus, slice modules are the cotilting modules with hereditary endomorphism rings, and actually a module with hereditary endomorphism ring is a tilting module if and only if it is a cotilting module.)

<u>PROPOSITION:</u> Let S be a slice in B-mod. Then $S = \langle {}_BS \rangle$ for a slice module ${}_BS$; in particular, B is a tilted algebra.

This shows that B is a tilted algebra if and only if B-mod contains a slice.

Slices are separating subcategories! More precisely, let S be a slice in B-mod, say $S = \langle {}_BS \rangle$, where ${}_BS$ is a slice module. Then S separates $C(\tau_B S)$ from $G(\tau_B^- S)$. Let us outline the proof. We have noted above that $(C({}_BS), G(\tau_B^- S))$ is a split torsion pair. Since the notion of a slice is self-dual, we see that similarly $(C(\tau_B S), G({}_BS))$ is a split torsion pair, and the definition of a slice implies that $G({}_BS) \cap C({}_BS) = S$, and also that $G(\tau_B^- S) \subseteq G({}_BS)$. It follows that $C(\tau_B S)$, S, and $G(\tau_B^- S)$ exhaust A-mod, and that the various zero-map conditions are satisfied. It remains to be seen that any map $f : P \longrightarrow Q$ with P in $C(\tau_B S)$, Q in $G(\tau_B^- S)$, factors through a module in S. Note that for any P in $C(\tau_B S)$, there exists an exact sequence $0 \to P \to S' \to S'' \to 0$, with both S', S'' in S (thus, S-codim $P \leq 1$). [Let ${}_BS = D({}_AT)$, where A is hereditary, ${}_AT$ a tilting module, and let $P = \mathrm{Hom}_A({}_AT_B, X)$, with X in $G({}_AT)$. Choose an injective resolution $0 \to X \to I' \to I'' \to 0$ of X, and apply $F = \mathrm{Hom}_A({}_AT_B, -)$. Since $\mathrm{Ext}^1_A({}_AT, X) = 0$, the sequence $0 \to FX \to FI' \to$ $\to FI'' \to 0$ still is exact. It remains to observe that both FI', FI'' belong to S.] With Q also $\tau^- Q$ belongs to $G(\tau_B^- S)$, thus $\mathrm{Hom}(\tau^- Q, S'') = 0$, therefore $\mathrm{Ext}^1(S'', Q) = 0$. But this implies that f can be factored through the given inclusion $P \longrightarrow S'$.

Assume now that B is a connected tilted algebra, and let S be a slice in B-mod. Since B is connected, the indecomposable modules in S all belong to a fixed component Γ, and we want to look at such components for a while. Let ${}_BS$ be a slice module, with $S = \langle {}_BS \rangle$, and we may assume that $A = \mathrm{End}({}_BS)$ is basic. Since A is hereditary, basic and connected, $A = k\Delta$ for some finite connected quiver Δ without oriented cycles. Now, S is equivalent to the category A-inj of all injective A-modules, and A-inj can be identified with the path category of Δ. On the other hand, constructing inductively the τ-translates, and the τ^--translates of the indecomposable direct summands of S, we see that they exhaust the component Γ. It follows that Γ is isomorphic to a full translation subquiver of $\mathbb{Z}\Delta$. Also, if we denote by $\overline{S}$ the module class whose indecomposable modules are those belonging to Γ, then it is easy to see that $\overline{S}$ is again a separating subcategory.

A component Γ containing the indecomposable modules of a slice S will be called a connecting component. The reason is the following: Let $S = <_BS>$, $\mathrm{End}(_BS) = A = k\Delta$, $_AT = D(S_A)$, and consider the functors $F = \mathrm{Hom}_A(_AT_B, -)$ and $F' = \mathrm{Ext}^1_A(_AT_B, -)$. The preinjective A-modules belonging to $G(_AT)$ go under F to modules in $\overline{S} \cap C(_BS)$, the preprojective A-modules which belong to $C(\tau_AT)$ go under F' to modules in $\overline{S} \cap G(\tau_B^-S)$, thus $\overline{S}$, in some sense, connects the preinjective and the preprojective component of A-mod:

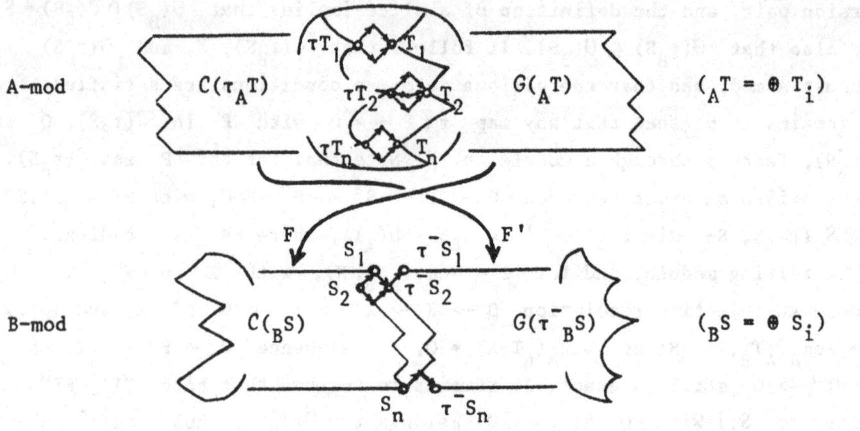

Actually, in case (A is representation-infinite and) no non-zero direct summand of $_AT$ is preinjective, then all preinjective A-modules belong to $G(_AT)$, and F gives an equivalence between the full subcategory of all preinjective A-modules, and $\overline{S} \cap C(_BS)$. Similarly, if (A is representation-infinite and) no non-zero direct summand of $_AT$ is preprojective, then all preprojective A-modules belong to $C(\tau_AT)$, and F' gives an equivalence between the full subcategory of all preprojective A-modules and $\overline{S} \cap G(\tau_B^-S)$. In particular, for $_AT$ regular, Γ is isomorphic to $\mathbb{Z}\Delta$, thus regular. On the other hand, if $_AT$ is not regular, then Γ cannot be a regular component. (For, if $_AT$ has an indecomposable preinjective direct summand, say T_i, then $[FT_i]$ will be a projective vertex of Γ. Similarly, assume that T_j is an indecomposable preprojective direct summand of T; if T_j is not projective, then $[F'(\tau_AT_j)]$ is an injective vertex of Γ, if T_j is projective, then $[F(\nu T_j)]$ is an injective vertex of Γ, where νT_j is the injective envelope of $T_j / \text{rad } T_j$).

In order to show the existence of components of the form $\mathbb{Z}\Delta$, it is sufficient to exhibit a regular tilting A-module, where $A = k\Delta$.

<u>PROPOSITION:</u> Let A be a finite-dimensional connected hereditary algebra, say $A = k\Delta$ for a quiver Δ. There exists a regular tilting A-module if and only if Δ is neither of Dynkin nor of Euclidean type, and has more than two vertices.

One direction of the proof is rather easy: In case Δ is of Dynkin or Euclidean type, one may prove without difficulties that $\mathbb{Z}\Delta$ does not allow an unbounded additive function, thus it is impossible to have a component of an Auslander-Reiten quiver of the form $\mathbb{Z}\Delta$. For $A = k\Delta$, where Δ has precisely two vertices, (and at least two arrows), any non-zero regular module X can be shown to satisfy $\mathrm{Ext}^1(X,X) \neq 0$, thus, also in this case, there cannot exist a regular tilting module. The converse implication is shown by constructing effectively regular tilting modules, see [Ri 3]. The argument shows that there cannot exist any algebra at all with a component of the form $\mathbb{Z}\Delta$, with Δ of Dynkin of Euclidean type. Since it is known that the Auslander-Reiten quiver of an algebra never contains a cyclic sectional path [BS], we obtain the following Corollary:

<u>COROLLARY:</u> Let Δ be a finite connected quiver, and $|\Delta_0| \neq 2$. Then, $\mathbb{Z}\Delta$ can be realised as a component of the Auslander-Reiten quiver Γ_A of some algebra A if and only if Δ has no oriented cycle and is neither of Dynkin nor of Euclidean type.

<u>Problem 7.</u> Let Δ be a quiver with two vertices, and at least three arrows, but no oriented cycles. Is it possible to realise $\mathbb{Z}\Delta$ as a component of some Γ_A?

We return to the investigation of sincere directing modules, say let M be a sincere directing B-module. Denote by $\mathcal{S}(M\to)$ the module class in B-mod such that an indecomposable B-module X belongs to $\mathcal{S}(M\to)$ if and only if first $M \preceq X$, and second, there does not exist an indecomposable non-projective B-module Z with both $M \preceq \tau Z$ and $Z \preceq X$. Similarly, denote by $\mathcal{S}(\to M)$ the module class in B-mod such that an indecomposable B-module Y belongs to $\mathcal{S}(\to M)$ if and only if $Y \preceq M$, and no indecomposable non-projective B-module Z satisfies both $Y \preceq \tau Z$ and $Z \preceq M$. Then one may show quite easily that both $\mathcal{S}(M\to)$ and $\mathcal{S}(\to M)$ are slices. In particular, B is a tilted algebra. Let $\mathcal{P}$, $\overline{\mathcal{S}}$, $\mathcal{Q}$ be the module classes in B-mod with $\overline{S}$ separating $\mathcal{P}$ from $\mathcal{Q}$, and such that the indecomposable modules in $\overline{S}$ are those which belong to the same component as $S(M\to)$ and $S(\to M)$. If we realize $S(\to M)$ as the set of images of the injective A-modules under a functor $F: \mathrm{Hom}_A({}_AT_B,-)$, where A is hereditary, ${}_AT$ a tilting module, and $B = \mathrm{End}({}_AT)$, we see that $N \preceq M$ for all indecomposable B-modules N in the image of F. This shows that $N \preceq M$ for all indecomposable modules in $\mathcal{P}$. Similarly, $M \preceq N$ for all indecomposable modules in $\mathcal{Q}$. It follows that the indecomposable modules N with both $N \npreceq M$ and $M \npreceq N$ belong to $\overline{S}$. Let $A = k\Delta$, where Δ is a quiver with n vertices. The component of Γ_B given by the modules in $\overline{S}$ embeds into $\mathbb{Z}\Delta$, and the number of vertices in $\mathbb{Z}\Delta$ which are incomparable with a fixed vertex is bounded by $\frac{n(n-1)}{2}$. This shows that there are at most $\frac{n(n-1)}{2}$ indecomposable modules N satisfying both $N \preceq M$ and $M \preceq N$.

Let us stress that any indecomposable module in a connecting component is directing; in particular, the regular connecting components are regular components which only contain directing modules!

As we have seen, the structure of a connecting component is known, at least if it is a regular component. Of course, we are also interested in the remaining components of a tilted algebra.

<u>PROPOSITION.</u> A regular component of a tilted algebra which is not a connecting component is either a regular tube or of the form $\mathbb{Z}A_\infty$.

<u>Problem 8.</u> What are the possible structures of non-regular components of tilted algebras?

Of special interest are the endomorphism rings $B = \mathrm{End}({}_AT)$, where A again is hereditary, connected, and representation-infinite, and ${}_AT$ is a tilting module which is, in addition, preprojective. In this case, B will be called a <u>concealed</u> algebra. Note that for a preprojective tilting module ${}_AT$ (with A hereditary), all but only finitely many indecomposable A-modules belong to $\mathcal{G}({}_AT)$, and therefore all but finitely many indecomposable B-modules belong to $\mathcal{C}({}_BS)$, (where $B = \mathrm{End}({}_AT)$ and ${}_BS = D(T_B)$).

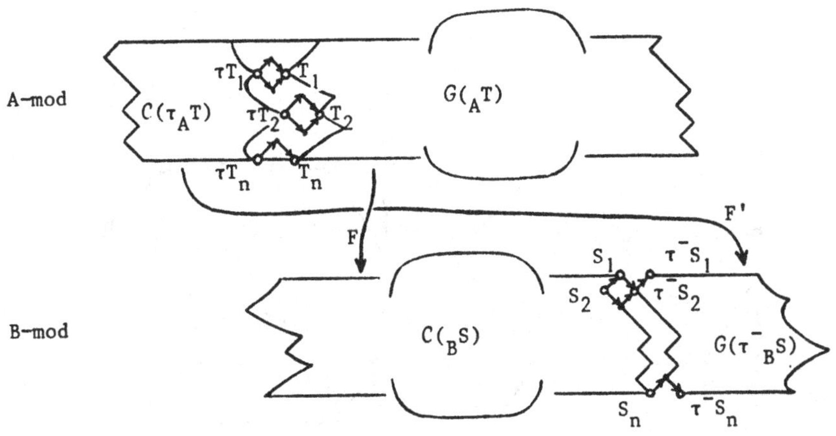

We see that a concealed algebra B has two non-regular components, one is a preprojective component (the images of the indecomposable preprojective modules in $G(_A T)$ under F), the other a preinjective component (the modules which are either images under F of the indecomposable preinjective A-modules or images under F' of the indecomposable preprojective modules in $C(\tau_A T)$), and the remaining components of Γ_B correspond, under F, to the regular components of Γ_A. Thus, the module categories of concealed algebras look like "concealments" of the module categories of hereditary algebras.

The tame concealed algebras play a particular role in representation theory, they are the minimal representation-infinite algebras A which have a preprojective component. Of course, an algebra A is said to be minimal representation-infinite provided A itself is representation-infinite, whereas all proper factor algebras of A are representation-finite. Also, it is not difficult to see that any connected, representation-infinite algebra C with a preprojective component has a factor algebra which is a tame concealed algebra. The tame concealed algebras have been classified by Happel and Vossieck, and the list of these algebras has been reprinted nearly everywhere, so we content ourselves by mentioning only some of its features: these algebras may be divided into different types, according to the type of the corresponding hereditary algebra. The only tame concealed algebras of type $\widetilde{\mathbb{A}}_n$ are the hereditary algebras of type $\widetilde{\mathbb{A}}_n$; there are four kinds of tame concealed algebras of type $\widetilde{\mathbb{D}}_n$:

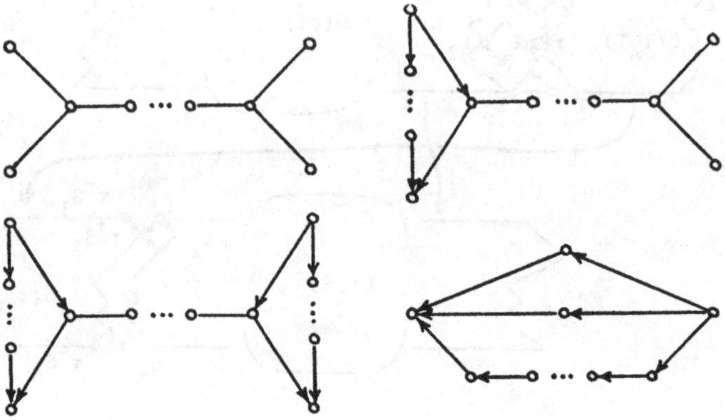

Here, the unoriented edges may be oriented arbitrarily, and, as relations, one has to take the sum of all paths from some vertex to another, provided there are at least two such paths. In addition, there are finitely many algebras of types $\widetilde{\mathbb{E}}_6$, $\widetilde{\mathbb{E}}_7$, $\widetilde{\mathbb{E}}_8$, given by quivers with 7, 8, or 9 vertices, respectively. The number of isomorphism classes of algebras of types $\widetilde{\mathbb{E}}_6$, $\widetilde{\mathbb{E}}_7$, $\widetilde{\mathbb{E}}_8$ is 56, 437, and 3809, respectively. One of the reasons for the general interest in the list of Happel-Vossieck is its usefulness for checking finite representation type; in particular, the socalled Bongartz criterion makes use of this list (so the list is sometimes also referred to as the bazar of Bongartz-Happel-Vossieck).

From the presentation above, one may have obtained a feeling of unsymmetry between the preprojective and the preinjective component of a concealed algebra B, but this is misleading: Also the preprojective component is a connecting component, since B can be written also as the endomorphism ring of a preinjective tilting module. In particular, we see that an algebra may have slices in two different components.

<u>PROPOSITION:</u> A (tilted) algebra B has at most two components containing slices, and if two, then B is a concealed algebra.

For the proof of this proposition, one needs additional techniques which we will report on in the third lecture. These techniques will shed more light on the connecting components and clarify the use of tilting functors.

Our next topic are the separating tubular families. Given an algebra A_0 and an A_0-module R_0, the <u>one-point extension</u> $A_0[R_0]$ of A_0 by R_0 is the matrix algebra $A_0[R_0] = \begin{bmatrix} A_0 & R_0 \\ 0 & k \end{bmatrix}$, its elements are the matrices $\begin{bmatrix} a & r \\ 0 & b \end{bmatrix}$ with $a \in A_0$, $b \in k$, $r \in R_0$, subject to the usual addition and multiplication of matrices. The $A_0[R_0]$-modules can be written as triples $X = (X_0, X_\omega, \gamma_X)$, where X_0 is an A_0-module, X_ω a k-space, and $\gamma_X : R_0 \otimes_k X_\omega \longrightarrow X_0$ is A_0-linear. The $A_0[R_0]$-module $E(\omega) = (0,k,0)$ is simple and injective, and conversely, any algebra with a simple injective module can be written as a one-point extension.

We consider an algebra A_0 with a sincere directing A_0-module W_0. In particular, A_0 has global dimension at most 2, and we denote by C_0 its Cartan matrix. As we have seen above, W_0 belongs to a slice, say $\langle S_o \rangle$, where S_0 is a slice A_0-module, and we can assume that its endomorphism ring is basic, thus of the form $k\Delta$. Note that the vertices of Δ correspond to the isomorphism classes of indecomposable summands of S_0; in particular, one of the vertices of Δ corresponds to $[W_0]$. We say that W_0 is a <u>wing</u> module of type $(n_1,\dots,n_t)$ provided the underlying graph of Δ is the star $\mathbb{T}_{n_1,\dots,n_t}$ and $[W_0]$ corresponds to the center of the star. (The star $\mathbb{T}_{n_1,\dots,n_t}$ is obtained from the disjoint union of copies of $\mathbb{A}_{n_i}$, $1 \le i \le t$, by choosing one endpoint in each $\mathbb{A}_{n_i}$, and identifying these endpoints to a single vertex, the center of the star). So assume now that W_0 is a wing A_0-module. The connecting component of A_0-mod containing W_0 has, in the vicinity of W_0, the following shape:

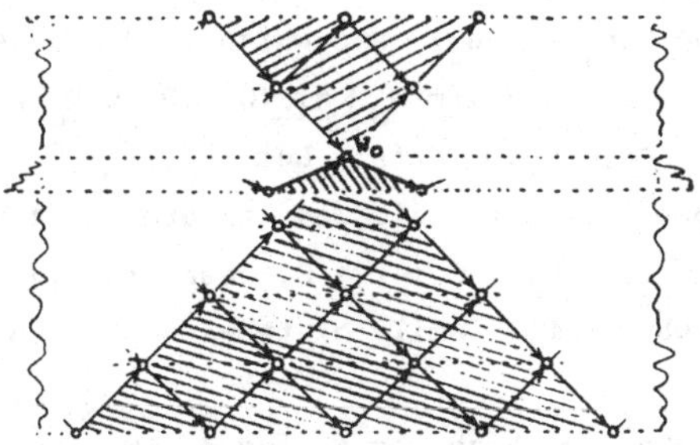

We have shaded the "wings" of W_0, these are given by those indecomposable modules which are successors of indecomposable modules in $S(\to W_0)$, and predecessors of indecomposable modules in $S(W_0 \to)$. We will say that W_0 is dominated by the A_0-module R_0 provided, first of all,

$$\underline{\dim}\, R_0 = (\underline{\dim}\, W_0)(I + C_0^{-1} C_0^T),$$

and second, for any $0 \neq \lambda : R_0 \longrightarrow W_0$, the $A_0[R_0]$-module $W(\lambda) = (W_0, k, \lambda)$ satisfies proj.dim.$W(\lambda) \leq 1$. It is easy to construct algebras with wing modules, but not every wing module is dominated by some module. In case W_0 is a wing A_0-module where A_0 is representation-finite, and $\underline{\dim}\, W_0$ is a maximal root, W_0 is always dominated by a projective module. Also other examples of dominated wing modules are known. In particular, the structure theory for the module category of a canonical algebra, as explained in the next lecture, rests on the use of certain dominated wing modules.

The separating subcategories which we are going to construct will be tubular families. By definition, a module class T is called a regular tubular $\mathbb{P}_1 k$-family of type $(n_1, \dots, n_t)$, provided T is equivalent to the mesh category $k(\Gamma)$, where Γ is the disjoint union of translation quivers $\Gamma(\rho)$, $\rho \in \mathbb{P}_1 k$, each $\Gamma(\rho)$ being a regular tube, and such that t of these tubes are of the form $\mathbb{Z}A_\infty / n_i$, $1 \leq i \leq t$, whereas the remaining ones all are homogeneous. Note that such a tubular family consists of various components $T(\rho)$, one for each $\rho \in \mathbb{P}_1 k$, with $T(\rho)$ equivalent to $k(\Gamma(\rho))$. We will say that T is a separating family, say separating P from Q, provided T is a separating subcategory, and, in addition, any map $f : P \longrightarrow Q$, with P in P, Q in Q may be factored through an object of any of the module classes $T(\rho)$, $\rho \in \mathbb{P}_1 k$. Now we may state the main theorem:

THEOREM. Let W_0 be a sincere directing A_0-module, which is a wing module of type $(n_1, \dots, n_t)$ and which is dominated by the A_0-module R_0. Let $A = A_0[R_0]$, and $w = \underline{\dim}\, W_o + \underline{\dim}(0,k,0) \in K_0(A)$. Denote by c_w the linear form $\iota_w = \langle w, - \rangle$ on $K_0(A)$. Let P_w, T_w, Q_w be the module classes in A-mod whose indecomposable modules are those indecomposable X which satisfy $\iota_w(\underline{\dim}\, X) < 0$, $= 0$, or > 0, respectively. Then T_w is a separating regular tubular $\mathbb{P}_1 k$-family of type $(n_1, \dots, n_t)$, separating P_w from Q_w.

Let us show in which way a wing of W_0 in A_0-mod gives rise to a tube in A-mod. We write $\langle -, - \rangle_0$ for the bilinear form and χ_0 for the quadratic form on $K_0(A)$. The condition $\underline{\dim}\, R_0 = (\underline{\dim}\, W_0)(I + C_0^{-1} C_0^T)$

implies that

$$\begin{aligned}\langle \underline{\dim} R_o, \underline{\dim} W_o \rangle_o &= (\underline{\dim} R_o)\, C_o^{-T} (\underline{\dim} W_o)^T \\ &= (\underline{\dim} W_o)(C_o^{-T} + C_o^{-1})(\underline{\dim} W_o)^T \\ &= 2\, \chi_o(\underline{\dim} W_o) = 2.\end{aligned}$$

Taking into account also the second condition on R_o, it follows quite easily that $\dim \mathrm{Hom}(R_o, W_o) = 2$. A wing of W_o in A_o-mod is of the following form:

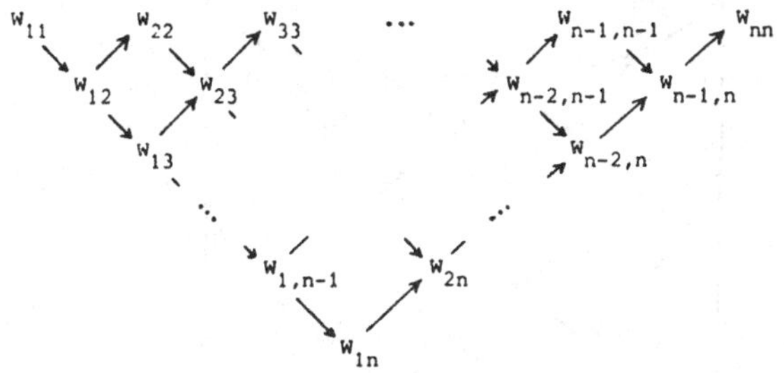

where all W_{ij} are indecomposable A_o-modules, $W_{1n} = W_o$, and the exhibited maps are irreducible. Again, using the domination conditions, one can show that $\dim \mathrm{Hom}(R_o, W_{ij}) = 1$ for $i = 0$, $1 \le j \le n-1$, and for $2 \le i \le n$, $j = n$, whereas $\mathrm{Hom}(R_o, W_{ij}) = 0$ for $2 \le i \le n-1$, $2 \le j \le n-1$. Up to scalar multiples, there is a unique non-zero map $\rho : R_o \longrightarrow W_o$ which factors through W_{11}. We may consider A_o-mod as a full subcategory of A-mod, identifying the A_o-module X_o with the triple $(X_o, 0, 0)$. Also, given an A_o-module Y_o, we write $\overline{Y}_o$ for the A-module $(Y_o, \mathrm{Hom}(R_o, Y_o), e)$, where $e : R_o \times \mathrm{Hom}(R, Y_o) \longrightarrow Y_o$ is the evaluation map. In A-mod, we obtain from the diagram above the following one:

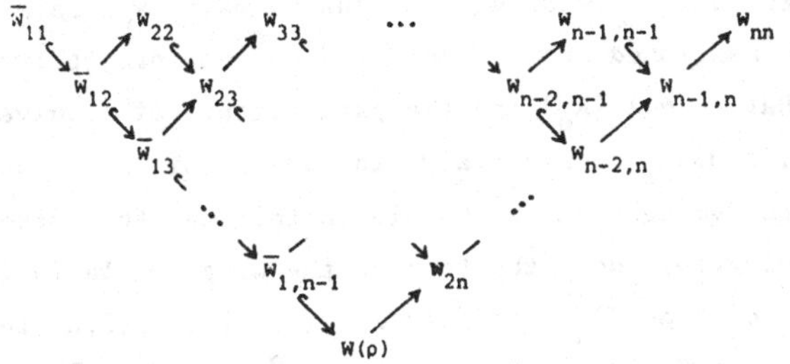

It is not difficult to see that the maps exhibited in the diagram again are irreducible, thus we look at part of a component $\mathcal{T}(\rho)$ of A-mod. A calculation shows that the dimension vector of the Auslander-Reiten translate $\tau_A\overline{W}_{11}$ is just $\underline{\dim}\, W_{nn}$. Since W_{nn} is uniquely determined by its dimension vector, this actually shows $\tau_A\overline{W}_{11} = W_{nn}$, thus W_{nn} is a τ-periodic module of period n. In this way, the component $\mathcal{T}(\rho)$ turns out to be a regular tube of the form $\mathbb{Z}\mathbb{A}_\infty / n$.

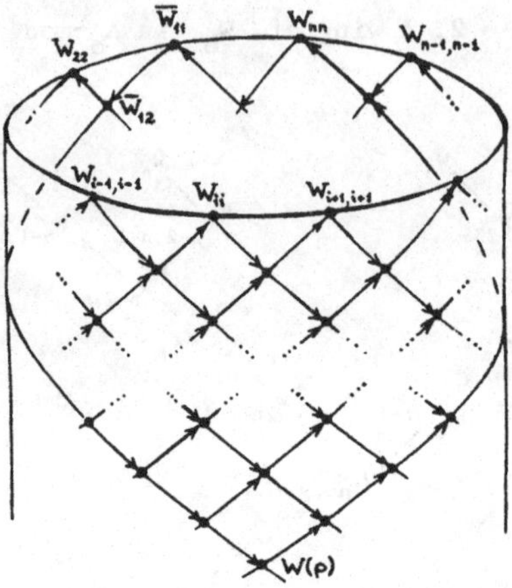

For any $0 \neq \rho \in \mathrm{Hom}(R_0,W_0)$, let $\mathcal{T}(\rho)$ be the component which contains $W(\rho)$. Then $\mathcal{T}(\rho)$ is a regular tube and $\mathcal{T}(\rho)$ is homogeneous if and only if ρ does not factor through an irreducible map $Y_0 \longrightarrow W_0$, with Y_0 indecomposable. Also, $\mathcal{T}(\rho) = \mathcal{T}(\rho')$ if and only if $\rho \sim \rho'$ in $\mathbb{P}\mathrm{Hom}(R_0,W_0)$. Thus, the index set for the tubular family constructed in this way is $\mathbb{P}\,\mathrm{Hom}\,(R_0,W_0) = \mathbb{P}_1 k$.

As a typical application, consider the case of $A_0 = k\Delta^0$, where Δ^0 is a quiver of type $\mathbb{A}_n$, $\mathbb{D}_n$, $\mathbb{E}_6$, $\mathbb{E}_7$, or $\mathbb{E}_8$, and W_0 the unique maximal indecomposable A_0-module. Note that always W_0 is a wing module, and, as we have mentioned above, dominated by some projective module R_0. It turns out that $A = A_0[R_0]$ is the path algebra of a quiver of type $\tilde{\mathbb{A}}_n$, $\tilde{\mathbb{D}}_n$, $\tilde{\mathbb{E}}_6$, $\tilde{\mathbb{E}}_7$, or $\tilde{\mathbb{E}}_8$, respectively (the quiver of A is obtained from Δ^0 by adding one vertex, and we obtain in this way the corresponding extended Dynkin diagram). Now, the type of the wing module is (2,2,n-2), in case Δ^0 is of type $\mathbb{D}_n$, and (3,3,2), (4,3,2), (5,3,2) in case Δ is of type $\mathbb{E}_6$, $\mathbb{E}_7$, or $\mathbb{E}_8$, respectively. For Δ^0 of type $\mathbb{A}_n$, the type of W_0 is of the form (p,q), with $p+q = n+1$, and depends on the orientation of Δ^0. In all cases, it follows that A has a regular tubular family $\mathcal{T} = \mathcal{T}_w$ of the same type, and $\mathcal{T}$ comprises just all regular modules. The modules in $\mathcal{P} = \mathcal{P}_w$ are the preprojective ones, those in $\mathcal{Q} = \mathcal{Q}_w$ the preinjective ones. In this way, we obtain the complete structure of the module category for any tame hereditary algebra. Any regular tubular family is, as a category in its own right, an abelian category which is serial (this means that any indecomposable object has a unique chain of subobjects). In particular, the category of all regular modules over a tame hereditary algebra is a serial abelian category.

There are also other cases of one-point extensions $A = A_o[R_o]$, where a separating subcategory $\mathcal{S}_o$ of A_o-mod gives rise to a separating subcategory $\mathcal{S}$ of A-mod. Assume that there is given a subcategory $\mathcal{S}_o$ of A_o-mod which separates $\mathcal{P}_o$ from $\mathcal{Q}_o$. The first case to be mentioned is that of R_o belonging to $\mathcal{Q}_o$. In this case, $\mathcal{S}_o$, considered as a subcategory of A-mod, is still separating: denote by $\mathcal{Q}$ the full subcategory of A-mod of all triples $(X_o, X_\omega, \gamma_X)$ with X_o in $\mathcal{Q}$. Then $\mathcal{S}_o$ separates $\mathcal{P}_o$ from $\mathcal{Q}$. The reason is that $\mathrm{Hom}(R_o, M_o) = 0$ for all M_o in $\mathcal{P}_o$ or $\mathcal{S}_o$. Here, two of the module classes $\mathcal{P}_o$, $\mathcal{S}_o$, $\mathcal{Q}_o$ are not changed at all when going from A_o-mod to A-mod, however one should observe that the change of $\mathcal{Q}_o$ to $\mathcal{Q}$ may be drastically. There is a second, less trivial case, the one-point extension $A = A_o[R_o]$, where R_o is a ray module in a component which is contained in $\mathcal{S}_o$.

A path $x_1 \longrightarrow x_2 \longrightarrow \dots \longrightarrow x_{n-1} \longrightarrow x_n$ in a translation quiver Γ is said to be <u>sectional</u> provided $\tau x_{i+1} \neq x_{i-1}$ for all possible i. A vertex v in Γ is said to be a <u>ray vertex</u> provided for any $i \in \mathbb{N}$, there is precisely one sectional path of length i and starting in v and such that the endpoints of the sectional paths of different length are different. Thus, given a ray vertex v, we obtain a unique infinite sectional path

$$v = v[1] \longrightarrow v[2] \longrightarrow \dots \longrightarrow v[i] \longrightarrow \dots$$

with pairwise different vertices $v[i]$, this path is called the <u>ray</u> starting at v. An indecomposable module V will be called a <u>ray module</u> provided the component Γ to which V belongs, is standard, and $[V]$ is a ray vertex in Γ. If V is a ray module, the indecomposable module M with $[M] = [V][i]$ will be denoted by $V[i]$.

Let V_o be a ray A_o-module and let Γ^o be the component to which V_o belongs. Let $A = A_o[V_o]$, and denote by Γ the component of Γ_A to which V_o, considered as an A-module, belongs. Then Γ contains all of Γ^o and, in addition, the modules $\overline{V_o[i]}$. The module $\overline{V}_o$ is again a ray module, and the ray starting in $[\overline{V}_o]$ is given by

$$[\overline{V}_o] \longrightarrow [\overline{V_o[2]}] \longrightarrow \dots \longrightarrow [\overline{V_o[i]}] \longrightarrow \dots .$$

Since this ray comprises just those indecomposable modules which belong to Γ and not to Γ^o, we say that Γ is obtained from Γ^o by <u>inserting a ray</u>. Assume now that V_o belongs to a separating subcategory S_o, say separating P_o from Q_o, and that, in addition, $\mathrm{Hom}(V_o,S_o) = 0$ for all indecomposable A_0-modules S_o in S_o which do not belong to Γ^o. This last condition is always satisfied in case S_o is standard. Denote by S the module class in A-mod which is obtained from S_o by adding the modules $\overline{V_o[i]}$ (and closing under direct sums), let Q be the set of all A-modules (X_o,X_ω,γ_X), with X_o having no direct summand in P_o or of the form $V_o[i]$. Then S is a separating subcategory, it separates P_o from Q.

These last considerations may be applied to the case of A_0 being a tame hereditary algebra, and V_o a regular ray A_0-module. We recall that the regular modules of a tame hereditary algebra form a regular tubular family. Now, the ray vertices of a regular tube $\mathbb{Z}A_\infty/n$ are those vertices which belong to the "mouth"; these are those vertices of $\mathbb{Z}A_\infty/n$ which are end points of precisely one arrow (and starting points of precisely one arrow), thus $\mathbb{Z}A_\infty/n$ has precisely n ray vertices. Recall that the category of all regular A_0-modules is abelian, and the simple objects in this category are just the ray modules.

LECTURE 3
REPETITIVE ALGEBRAS

The third lecture is devoted to recent investigations of Happel. As we have seen in the previous lectures, the regular components of the Auslander-Reiten quiver of an algebra are usually rather well-behaved, whereas there are difficulties in dealing with non-regular components. Of course, given any translation quiver Γ, one may delete the τ-orbits which contain projective or injective vertices, in order to obtain a better behaved translation quiver Γ_s. Thus, instead of considering the Auslander-Reiten quiver Γ_A, it has been customary to consider the corresponding <u>stable Auslander-Reiten quiver</u> $(\Gamma_A)_s$, and to try to recover Γ_A from $(\Gamma_A)_s$. For example, Riedtmann's classification of the representation-finite selfinjective algebras was done along these lines. Of course, for a selfinjective algebra, the only vertices of Γ_A which do not belong to $(\Gamma_A)_s$ are those given by the indecomposable projective-injective modules, so there are only few such vertices, and their positions can be uniquely determined from the knowledge of $(\Gamma_A)_s$ and the restriction of the usual length function to $(\Gamma_A)_s$. On the other hand, there may be whole components Γ of Γ_A such that any τ-orbit in Γ contains a projective or an injective vertex, thus Γ_s is empty in this case, so it is impossible to recover Γ from Γ_s. For example, this is the case for any preprojective or preinjective component, but also for any component obtained from a regular tube by inserting rays. The latter example shows that there may be more subtle ways of deleting vertices in order to obtain a regular translation quiver from a non-regular one, however it always will be awkward to reinsert vertices. Let us outline a completely different strategy of investigating A-mod by considering solely regular components.

We will consider in the sequel (associative) algebras R (defined over k) which are not necessarily finite-dimensional and which are <u>not</u> required to have a unit element, however, we do require that

$R^2 = R$ (here, R^2 denotes the subspace of R generated by all products r_1r_2, with $r_1,r_2 \in R$). The algebras which we will encounter will at least have sufficiently many idempotents, since they may be thought of as being given by small preadditive categories. Any small k-preadditive category A gives rise to an algebra $\oplus A$, with underlying vectorspace $\oplus_{x,y} \mathrm{Hom}_A(x,y)$, where x,y range over all objects of A, and where the multiplication is given by the composition of maps in A whenever defined, and zero otherwise. Of course, in $\oplus A$ there are many idempotents, namely all the identity maps of the objects of A. However, $\oplus A$ has a unit element only in case A has only finitely many objects. Note that an algebra R is of the form $\oplus A$ if and only if there is a complete set $\{e_x \mid x \in I\}$ of pairwise orthogonal idempotents in R (the completeness means that $R = \oplus_{x,y} e_xRe_y$).

Given an algebra R, a (left) R-module M is always supposed to satisfy the condition $RM = M$ (as above, RM denotes the subspace of M generated by all elements of the form rm, with $r \in R$, $m \in M$). Note that in case R has a unit element, the condition $RM = M$ is equivalent to the usual one on M to be unital. If $\{e_x \mid x \in I\}$ is a complete set of pairwise orthogonal idempotents, and M is an R-module, then, as a vectorspace, M decomposes in the form $M = \oplus e_xM$. An R-module M is said to be finitely generated in case there are elements $m_1,\ldots,m_n \in M$ with $M = \Sigma Rm_i$. [Note that in general ${}_RR$ itself is not finitely generated! If R has a complete set of pairwise orthogonal idempotents and ${}_RR$ is finitely generated, then R actually has a unit element.] We denote by R-Mod the category of all R-modules, by R-mod the full subcategory of all finitely generated ones. If A is a small k-preadditive category, then $\oplus A$-Mod is nothing else than the category of additive contravariant functors from $\oplus A$ into the category k-Mod of k-vectorspaces, and $\oplus A$-mod is just the full subcategory of all finitely generated functors. [If $F : A \longrightarrow$ k-Mod is a contravariant functor, the corresponding $\oplus A$-module is given by the vectorspace $\oplus_x F(x)$, where x ranges over all objects x of A, and $\oplus A$ operates on this vectorspace as follows: for $\alpha \in \mathrm{Hom}_A(x,y)$, $m \in F(z)$, let $\alpha m = F(\alpha)(m)$ in case $y = z$, and $= 0$ otherwise. Conversely, given an $\oplus A$-module M, the corresponding functor is defined as follows: any object x of A is sent to 1_xM, any map $\alpha : x \longrightarrow y$ to the map $1_yM \longrightarrow 1_xM$ given by left multiplication by α.]

The algebra R is said to be <u>locally bounded</u> provided there exists a complete set of pairwise orthogonal primitive idempotents $\{e_x \mid x \in I\}$ with Re_x and e_xR being finite-dimensional, for all $x \in I$. Assume that R is locally bounded, and let $\{e_x \mid x \in I\}$ be a complete set of pairwise orthogonal primitive idempotents in R. The R-module Re_x will be denoted by P(x). One obtains in this way all possible indecomposable projective R-modules (there may be some dublications: some of the P(x) may be isomorphic). Similarly, let $Q(x) = \mathrm{Hom}_k(e_xR,k)$, one obtains in this way all possible indecomposable injective R-modules (again, may be, with dublications). For a locally bounded algebra R, any finitely generated R-module is of finite length, thus R-mod is abelian, and the simple R-modules are of the form P(x)/rad P(x). Also in this case, any finitely generated R-module has both a projective cover and an injective envelope in R-mod. It follows that R-mod has Auslander-Reiten sequences, and we will denote by Γ_R the corresponding Auslander-Reiten quiver. From now on, given a locally bounded algebra R, the R-modules we will deal with, always will be supposed to be finitely generated. We call an algebra R a <u>Frobenius-algebra</u> provided it is locally bounded, and the indecomposable projective R-modules coincide with the indecomposable injective R-modules. (Observe that we deviate from the usual terminology, even in case R being finite dimensional; the usual name in this case is that R is "Quasi-Frobenius", or selfinjective; whereas Nakayama's "Frobenius algebras" are finite-dimensional Frobenius-algebras which satisfy some additional multiplicity condition).

Given a Frobenius algebra R, we denote by R-<u>mod</u> the <u>stable</u> module category: its objects are the same as those of R-mod, namely the finitely generated R-modules, and given two finitely generated R-modules X,Y, the set of morphisms from X to Y in R-<u>mod</u> is denoted by <u>Hom</u>(X,Y), and <u>Hom</u>(X,Y) = $\mathrm{Hom}_R(X,Y)/\sim$, where $f \sim g$ iff $f - g$ factors through a projective R-module. The residue class of a map $f : X \longrightarrow Y$ in <u>Hom</u>(X,Y) will be denoted by $\underline{f}$. Note that the isomorphism classes of indecomposable objects in R-<u>mod</u> correspond naturally to the isomorphism classes of the indecomposable non-projective R-modules, thus to the vertices of the stable Auslander-Reiten quiver $(\Gamma_A)_s$. The stable module category R-<u>mod</u> preserves much information concerning R-mod.

<u>Lemma</u>. Let R be a Frobenius algebra, and X,Y indecomposable non-projective R-modules satisfying $\underline{\mathrm{Hom}}_R(X,Y) \neq 0$. Then there exists an indecomposable (non-projective) R-module M such that $\underline{\mathrm{Hom}}_R(X,M) \neq 0$, $\underline{\mathrm{Hom}}_R(M,Y) \neq 0$.

<u>Proof</u> (Vossieck): First of all, assume there exists $f : X \longrightarrow Y$ which is epi. We claim that $\underline{f}$ itself is non-zero. For assume $\underline{f} = \underline{f_1 f_2}$, where $f_1 : X \longrightarrow P$, $f_2 : P \longrightarrow Y$ with P projective. Since f is epi, and P is projective, there exists $f_2' : P \longrightarrow X$ with $f_2' f = f_2$, and therefore $f = f_1 f_2 = (f_1 f_2') f = (f_1 f_2')^n f$ for arbitrary $n \in \mathbb{N}$. Since f_1 is not split mono, and X is indecomposable, it follows that $(f_1 f_2')^n = 0$ for large n, thus $f = 0$, a contradiction. Similarly, we see that any mono map $g : X \rightarrow Y$ satisfies $\underline{g} = 0$.

In general, take an arbitrary non-zero map $h : X \longrightarrow Y$, and let M be an indecomposable direct summand of the image of h. Let $f : X \rightarrow$ and $g : M \longrightarrow Y$ be the maps induced by h, thus f is epi, g is mono. As a consequence, we see $\underline{\mathrm{Hom}}_R(X,M) \neq 0$, $\underline{\mathrm{Hom}}_R(M,Y) \neq 0$.

It follows that for a Frobenius algebra R, a search for directing modules and for separating subcategories in R-mod can be carried out in R-$\underline{\text{mod}}$. We will see below examples where we may use this technique.

Also, we should mention that for a general Frobenius algebra R, and Γ a component of Γ_R, the stable subquiver Γ_s is still connected. Of course, there is the following trivial exception: if R is a simple artinian ring, then Γ_R consists of a single vertex and no arrow, whereas $(\Gamma_R)_s$ is empty. [Also note the following slightly pathological case: Assume that R has an indecomposable projective R-module of length 2. If R is connected, then all the indecomposable projective R-modules are of length 2, the remaining indecomposable R-modules all are simple, and $(\Gamma_R)_s = \mathbb{Z}\mathbb{A}_1$ or $\mathbb{Z}\mathbb{A}_1/n$. Now, $\mathbb{Z}\mathbb{A}_1$ and $\mathbb{Z}\mathbb{A}_1/n$ are connected as translation quivers, but the only case which is connected as a mere quiver is $\mathbb{Z}\mathbb{A}_1/1$.

Of course, if R has no projective R-modules of length 2, and Γ is any component of Γ_R, then Γ_S will be connected not only as a translation quiver, but as a mere quiver.]

Given any finite dimensional algebra A, let us construct the corresponding <u>repetitive</u> algebra $\hat{A}$, as proposed by Hughes and Waschbüsch [HW]. It will be a Frobenius algebra and always infinite-dimensional (except in the trivial case $A = 0$ which we may exclude). Denote by Q the A-A-bimodule $Q = \mathrm{Hom}_k(A,k)$ (the bimodule actions are defined in the obvious way: given $a',a'' \in A$, $\varphi \in Q$, then $a'\varphi a''$ is the k-linear map which sends $a \in A$ to $\varphi(a''aa')$. The underlying vectorspace of $\hat{A}$ is given by

$$\hat{A} = \left(\bigoplus_{i \in \mathbb{Z}} A\right) \oplus \left(\bigoplus_{i \in \mathbb{Z}} Q\right),$$

we denote the elements of $\hat{A}$ by $(a_i,q_i)_i$, where $a_i \in A$, $q_i \in Q$, of course with almost all a_i,q_i being zero. The multiplication is defined by

$$(a_i,q_i)_i(a'_i,q'_i)_i = (a_ia'_i, a_{i+1}q'_i + q_ia'_i)_i.$$

We also may use the categorical description $\hat{A} = \oplus\, \hat{\mathcal{A}}$, which is derived as follows: Choose a complete set $\{e_x \mid x \in I\}$ of pairwise orthogonal primitive idempotents in A (of course, I is a finite set), let $\mathcal{A}$ be the category with object set I, with $\mathrm{Hom}_{\mathcal{A}}(x,y) = e_xAe_y$, for all $x,y \in I$, and with composition of morphisms being the multiplication in A. Thus, $A = \oplus\, \mathcal{A}$. Define the category $\hat{\mathcal{A}}$ as follows: as object set we take $I \times \mathbb{Z}$, instead of (x,n), where $x \in I$, $n \in \mathbb{Z}$, we write $x\langle n\rangle$, we define

$$\mathrm{Hom}_{\hat{\mathcal{A}}}(x\langle n\rangle, y\langle m\rangle) = \begin{cases} e_xAe_y & m = n, \\ \mathrm{Hom}_k(e_yAe_x,k) \quad \text{for} & m = n-1, \\ 0 & m \neq n,n-1. \end{cases}$$

and the composition of maps is derived from the multiplication in A and the A-A-bimodule structure of Q. [More precisely, the composition of maps $x\langle n\rangle \to y\langle n\rangle \to z\langle n\rangle$ is given by the multiplication in A, the compostion $x\langle n\rangle \to y\langle n\rangle \to z\langle n-1\rangle$ is given by the canonical map $e_xAe_y \otimes \mathrm{Hom}_k(e_zAe_y,k) \longrightarrow \mathrm{Hom}(e_zAe_x,k)$ which sends $e_xae_y \otimes \varphi$ to the linear form $e_zbe_x \longmapsto \varphi(e_zae_xae_y)$; similarly, the composition

$x\langle n\rangle \longrightarrow y\langle n-1\rangle \longrightarrow z\langle n-1\rangle$ is given by the canonical map $\mathrm{Hom}_k(e_yAe_x,k) \otimes e_yAe_z \longrightarrow \mathrm{Hom}(e_zAe_x,k)$ which sends $\psi \otimes e_yae_z$ to the linear form $e_zbe_x \longmapsto \varphi(e_yae_zbe_x)$.] It follows that $\hat{A} = \oplus \hat{A}$. Note that there exists an infinite cyclic group of automorphisms of both $\hat{A}$ and $\hat{A}$, given by shifting the indices, we denote by ν a generator of this group, namely the automorphism of $\hat{A}$ which sends $(a_i,q_i)_i$ to $(a_i',q_i')_i$ with $a_i' = a_{i-1}, q_i' = q_{i-1}$, and also the corresponding automorphism of $\hat{A}$ which sends $x\langle n\rangle$ to $x\langle n+1\rangle$.

The $\hat{A}$-modules can be written in the following way: $M = (M_i,f_i)_i$, where the M_i are A-modules, all but finitely many being zero, the f_i are A-linear maps $f_i = f_i^M : Q \otimes_A M_i \longrightarrow M_{i+1}$ such that the conditions $(Q\otimes f_i)f_{i+1} = 0$ are satisfied for all i, where always $i \in \mathbb{Z}$. [To wit, given such an $M = (M_i,f_i)_i$, consider $\oplus M_i$ as an $\hat{A}$-module using the scalar multiplication $(a_i,q_i)_i(m_i)_i = (a_im_i+(q_{i-1} \otimes m_{i-1})f_{i-1})_i$, where $(a_i,q_i)_i \in \hat{A}$, and $(m_i)_i \in \oplus M_i$. For the converse, note that the element $1_j = (\delta_{ij},0)_i \in \hat{A}$ (where δ_{ij} is the Kronecker symbol) is an idempotent, and actually the family $\{1_j \mid j \in \mathbb{Z}\}$ is a complete set of pairwise orthogonal idempotents. Given an $\hat{A}$-module M, we decompose M with respect to this family, thus $M_j = 1_j \cdot M$.] It is easy to calculate the indecomposable projective modules. Given $x \in I$, $n \in \mathbb{Z}$, we obtain

$$P(x\langle n\rangle)_i = \begin{cases} P_A(x) & i = n \\ Q_A(x) & \text{for } i = n-1 \\ 0 & i \neq n,n-1 \end{cases}$$

and $f_n^{P(x\langle n\rangle)} : Q \otimes P_A(x) \longrightarrow Q_A(x)$ the canonical isomorphism. Of course, this $\hat{A}$-module is also the indecomposable injective module corresponding to the vertex $x\langle n-1\rangle$, thus

$$P(x\langle n\rangle) = Q(x\langle n-1\rangle).$$

In particular, we see in this way that $\hat{A}$ is a Frobenius algebra.

There are countably many obvious embeddings of A-mod into $\hat{A}$-mod, indexed over $\mathbb{Z}$. The $-\langle n\rangle$ embedding with index n will send M to $M\langle n\rangle$, where $M\langle n\rangle_i$ for $i = n$, and $= 0$ otherwise. Observe that the composition of any of these embeddings $-\langle n\rangle$ with the canonical functor $\hat{A}$-mod $\longrightarrow$ $\hat{A}$-<u>mod</u> is still a full embedding. [The reason is the following: Given A-modules M_1,M_2 and a map $f : M_1 \longrightarrow M_2$, and suppose the corresponding map $f\langle n\rangle : M_1\langle n\rangle \longrightarrow M_2\langle n\rangle$ factors through a projective module. This implies that $f\langle n\rangle$ actually factors through a projective cover P of $M_2\langle n\rangle$. The indecomposable projective summands of P are of the form $P(a\langle n\rangle) = Q(a\langle n-1\rangle)$, and clearly $\mathrm{Hom}(M_1\langle n\rangle, Q(a\langle n-1\rangle)) = 0$.] The embedding $-\langle 0\rangle$ will be called the canonical embedding, and we will identify A-mod with its image under $-\langle 0\rangle$. Note that in this way we have achieved a full embedding of A-mod into the rather well-behaved category $\hat{A}$-mod.

We should observe that the category $\hat{A}$-mod may be defined alternatively as the category of graded modules over some graded algebra. To wit, let $T(A)$ be the trivial extension algebra of A. The underlying vectorspace of $T(A)$ is $A \oplus Q$ (recall that $Q = \mathrm{Hom}_k(A,k)$), and the multiplication is given by

$$(a,q)(a',q') = (aa',aq'+qa')$$

for $a,a' \in A$; and $q,q' \in Q$. The algebra $T(A)$ with the displayed decomposition $T(A) = A \oplus Q$ is a $\mathbb{Z}$-graded algebra, where $A \oplus 0$ are the elements of degree 0, those of $0 \oplus Q$ the elements of degree 1. We denote by $T(A)$-grmod the category of finitely generated $\mathbb{Z}$-graded modules over $T(A)$ and morphisms of degree zero. Obviously, a finitely generated $\mathbb{Z}$-graded module is of the form (M_i, f_i), where the M_i are A-modules, all but finitely many being zero, and the f_i are A-linear maps $Q \otimes M_i \longrightarrow M_{i+1}$ satisfying $(Q \otimes f_i)f_{i+1} = 0$ for all i. Thus $T(A)\text{-grmod} = \hat{A}\text{-mod}$.

We want to consider one example in detail. First, let us introduce the canonical algebras $C(\lambda,p)$ where $\lambda = (\lambda_0,\lambda_1,\dots,\lambda_t)$ is an $(t+1)$-tupel of pairwise different elements in $\mathbb{P}_1k$ (and we may suppose $\lambda_0 = \infty$, $\lambda_1 = 0$), and $p = (p_0,p_1,\dots,p_t)$ an $(t+1)$-tupel of positive integers, and $t \geq 1$. Take the disjoint union of linearly ordered quivers of types $\mathbb{A}_{p_0+1},\dots,\mathbb{A}_{p_t+1}$, the arrows in the i-th quiver will all be denoted by the letter α_i, and identify all sinks to a single vertex, and identify all sources again to a single vertex. Thus, we deal with the following quiver $\Delta(\lambda,p)$

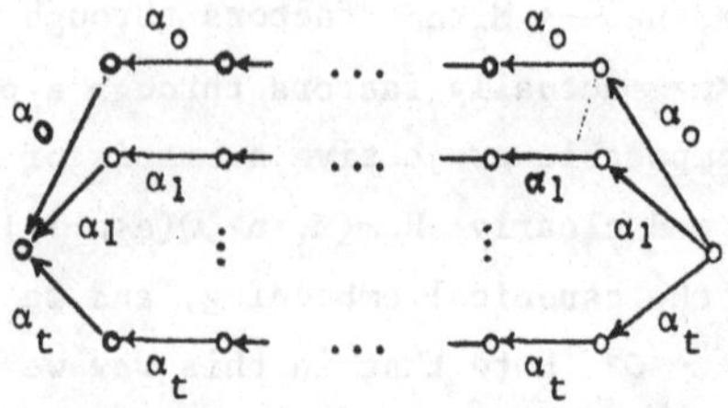

with paths from the source to the sink of length $p_o,\ldots,p_t$, respectively. The canonical algebra $C(\lambda,p)$ is the opposite of the path algebra of this quiver modulo the relations

$$\lambda_i\alpha_o^{p_o} + \alpha_1^{p_1} + \alpha_i^{p_i} = o, \quad \text{for } 2 \leq i \leq t.$$

[It may look fancy that we insist to deal with the opposite, since obviously the opposite of a canonical algebra is a canonical algebra again. The reason is that we want to identify the category $C(\lambda,p)$-mod with a category of representations of $\Delta(\lambda,p)$.] The Frobenius algebra $F(\lambda,p,q)$ which we want to consider is defined as follows: We start with an (n+1)-tupel $\lambda = (\infty,o,\lambda_2,\ldots,\lambda_t)$ of pairwise different elements of $\mathbb{P}_1k$, two (t+1)-tuples $p = (p_o,\ldots,p_t)$, $q = (q_o,\ldots,q_t)$ of positive integers, take countably many copies $\Delta\langle n\rangle$, $n \in \mathbb{Z}$ of the quiver $\Delta(\lambda,p)$, countably many copies $\Delta'\langle n\rangle$, $n \in \mathbb{Z}$ of the quiver $\Delta(\lambda,q)$, and identify the sink of $\Delta\langle n\rangle$ with the source of $\Delta'\langle n-1\rangle$ and call this vertex $a\langle n\rangle$, and the sink of $\Delta'\langle n\rangle$ with the source of $\Delta\langle n\rangle$ and call this vertex $b\langle n\rangle$. As relations on this quiver $\Delta(\lambda,p,q)$ we use first of all those which make all $\Delta\langle n\rangle$, $\Delta'\langle n\rangle$ into canonical algebras, and, in addition the following ones: Denote as above the arrows of any $\Delta\langle n\rangle$ by α_i $(o \leq i \leq t)$, and denote those of $\Delta'\langle n\rangle$ in the same way by β_i $(o \leq i \leq t)$. The additional relations are the following ones (whenever they make sense):

$$\alpha_i\beta_i = o, \quad \beta_i\alpha_i = o \qquad \text{for all } o \leq i \leq t,$$

$$\alpha_o^{p_o}\beta_1^{q_1} = \alpha_1^{p_1}\beta_o^{q_o}, \qquad \beta_o^{q_o}\alpha_1^{p_1} = \beta_1^{q_1}\alpha_o^{q_o}$$

$$\alpha_i^{u}\beta_j^{q_j}\alpha_i^{p_i-u+1} = 0 \qquad \text{for all } i \neq j,\ 1 \leq u \leq p_i$$

$$\alpha_i^{u}\beta_j^{q_j}\alpha_i^{p_i-u+1} = 0 \qquad \text{for all } i \neq j,\ 1 \leq u \leq q_i$$

and in this way, we obtain $F(\lambda,p,q)$ (agáin as the opposite of the path algebra modulo the listed relations).

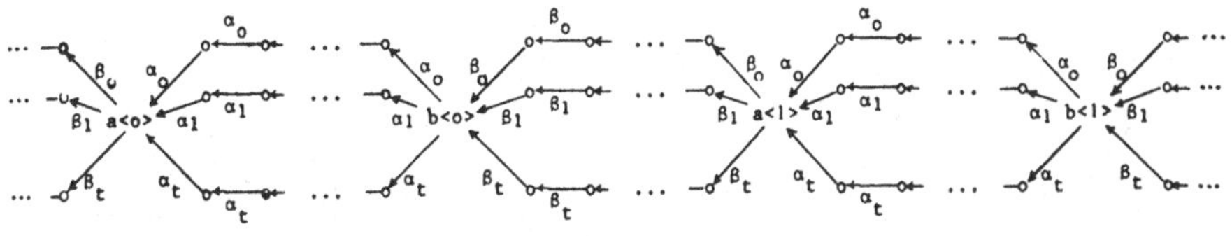

It is easy to check that $F(\lambda,p,q)$ is a Frobenius algebra, and it may be considered as the repetitive algebra $\hat{A}$ for many algebras A defined by full subquivers of $\Delta(\lambda,p,q)$ and the corresponding relations. For example, as the notation suggest, we may take the full subquiver of all vertices in $\Delta\langle o\rangle \cup \Delta'\langle o\rangle$ different from $a\langle 1\rangle$.

We are going to show in which way we can use the techniques presented in the last lecture in order to determine at least parts of the category $F(\lambda,p,q)$-mod, for arbitrary λ,p,q. First, we consider the canonical algebra $C(\lambda,p)$. Denote by $\mathcal{P}$ the additive subcategory of $C(\lambda,p)$-mod whose indecomposable objects are those representations of $\Delta(\lambda,p)$ for which all maps α_i are injective, and at least one α_i is not surjective, similarly, let $\mathcal{Q}$ be the additive subcategory of $C(\lambda,p)$-mod with indecomposables those representations for which all maps α_i are surjective and at least one α_i is not injective. Finally, let $\mathcal{T}$ be the additive subcategory whose indecomposables are the remaining ones (those for which either all α_i are bijective, or else some α_i is not injective and some α_j, may be the same, is not surjective). We observe that $\mathcal{T}$ is a separating tubular family of type p separating $\mathcal{P}$ from $\mathcal{Q}$. [For, we can apply the construction theorem for separating tubular families as presented in the last lecture: $C(\lambda,p)$ is the one-point extension of a hereditary algebra A_o with quiver a star, by a certain A_o-module R_o (the dimension vector of R_o is of the form $\begin{smallmatrix} & 11\ldots1 \\ 2 & 11\ldots1 \\ & 11\ldots1 \end{smallmatrix}$); the only non-serial indecomposable injective A_o-module is a wing A_o-module which is dominated by R_o, and it is not difficult to show that the module classes $\mathcal{P},\mathcal{T},\mathcal{Q}$ established there have the description given above.] Let us write down explicitely the mouth of one of the exceptional tubes $T(\rho)$ in T, say, the one with index $\rho = \infty$ and to be specific, in the case where $p_0 = 4$:

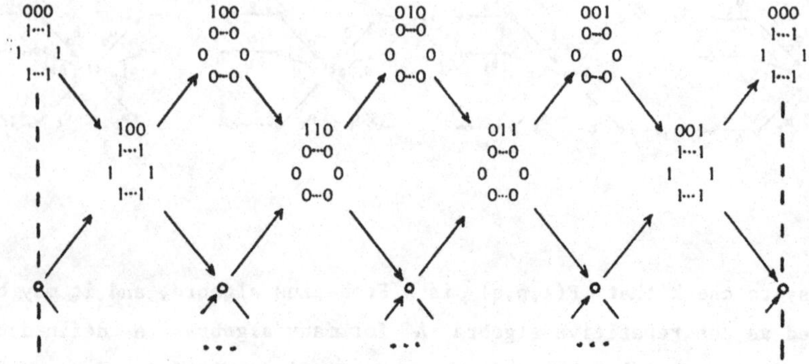

here, we have to identify the dashed vertical boundaries. What happens with this tubular family T when we consider $C(\lambda,p)$-mod as a full subcategory of $F(\lambda,p,q)$-mod, identifying $\Delta(\lambda,p)$ with $\Delta\langle o\rangle$?

First, consider the full subquiver

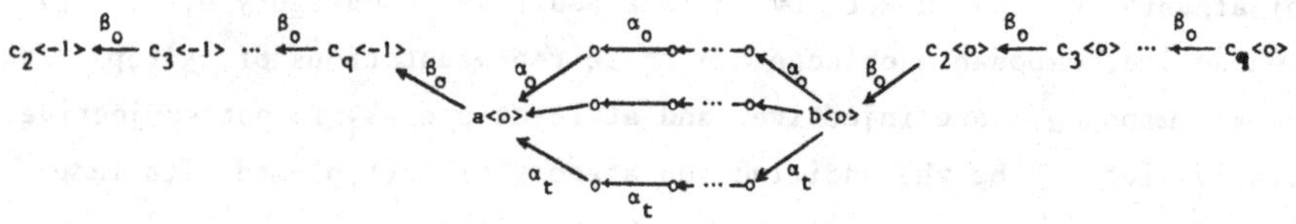

taking into account besides $\Delta\langle o\rangle$ also those arrows β_0 of $\Delta\langle -1\rangle$ which do not end in the sink of $\Delta'\langle -1\rangle$, and those arrows β_0 of $\Delta\langle o\rangle$ which do not start in the source of $\Delta\langle o\rangle$ (and the relations which live on this subquiver). We obtain this algebra from $C(\lambda,p)$ by successive tubular extensions, followed by successive tubular coextensions, and these extensions and coextensions change the tube $T(\infty)$ with index ∞, but leave untouched the remaining tubes in T. The new tube with index ∞ is obtained from $T(\infty)$ by first inserting $q-1$ rays, and then inserting the same number of corays; there are q_0-1 indecomposable modules which are both projective and injective, namely the modules $P(c_i\langle o\rangle) = Q(c_i\langle -1\rangle)$, $2 \le i \le q_0$. Deleting these modules, we obtain a stable tube of the form $\mathbb{Z}A_\infty/(p_0+q_0-1)$. Let us indicate the mouth of the new tube with index ∞

for the case $q_o = 3$

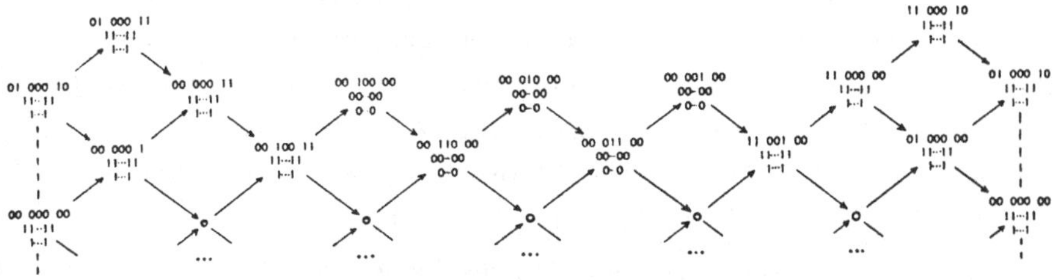

here again, the vertical dashed boundary lines have to be identified.

If we consider the full subquiver of $\Delta(\lambda,p,q)$ given by all vertices in $\Delta'\langle -1\rangle \cup \Delta\langle o\rangle \cup \Delta\ \langle o\rangle$ different from $b\langle -1\rangle$ and $a\langle 1\rangle$:

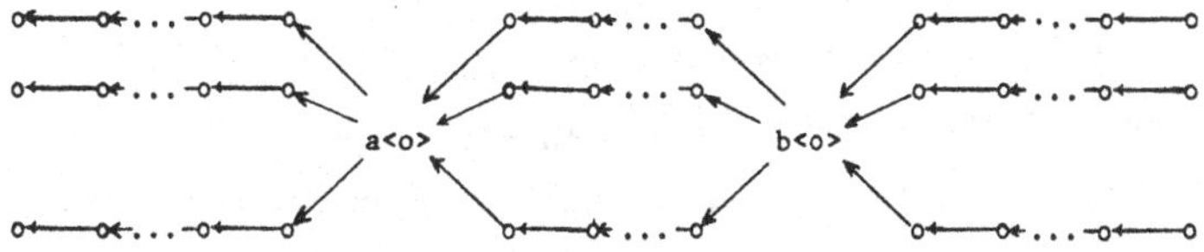

and all relations which live on this subquiver, we see that we obtain a tubular family $\mathcal{T}\langle o\rangle$ which contains the tubular family $\mathcal{T}$ of $C(\lambda,p)$ and which is obtained by inserting q_i-1 rays and also q_i-1 corays into the tube $\mathcal{T}(\lambda_i)$, $0 \le i \le t$. This tubular family $\mathcal{T}\langle o\rangle$ again is separating. If we continue to form one-point extensions and one-point coextensions in order to arrive finally at $F(\lambda,p,q)$, we see that the tubular family $\mathcal{T}\langle o\rangle$ remains unchanged, and still is separating. So denote by $\mathcal{P}\langle o\rangle$, $\mathcal{Q}\langle o\rangle$ the module classes in $F(\lambda,p,q)$ which are separated by $\mathcal{T}\langle o\rangle$. An $F(\lambda,p,q)$-module P belongs to $\mathcal{P}\langle o\rangle$ if and only if it satisfies the following two properties: First of all, P has to live on the union of all $\Delta'\langle 1\rangle$, $n \le -1$, and all $\Delta\langle m\rangle$, $m \le o$, and second, the indecomposable summands of the restriction of P to $\Delta\langle o\rangle$ all have to belong to $\mathcal{P}$.

What we have done starting with $\Delta\langle o\rangle$, we can do for any $\Delta\langle n\rangle$ and any $\Delta'\langle n\rangle$. We obtain tubular families $\mathcal{T}\langle n\rangle$ and $\mathcal{T}'\langle n\rangle$ whose stable type is $(p_o+q_o-1,\dots,p_t+q_t-1)$, and corresponding module classes $\mathcal{P}\langle n\rangle$, $\mathcal{Q}\langle n\rangle$ and $\mathcal{P}'\langle n\rangle$, $\mathcal{Q}'\langle n\rangle$. Let $\mathcal{M}\langle n\rangle = \mathcal{Q}\langle n\rangle \cap \mathcal{P}'\langle n\rangle$, and $\mathcal{M}'\langle n\rangle = \mathcal{Q}'\langle n\rangle \cap \mathcal{P}\langle n+1\rangle$. Then the following module classes

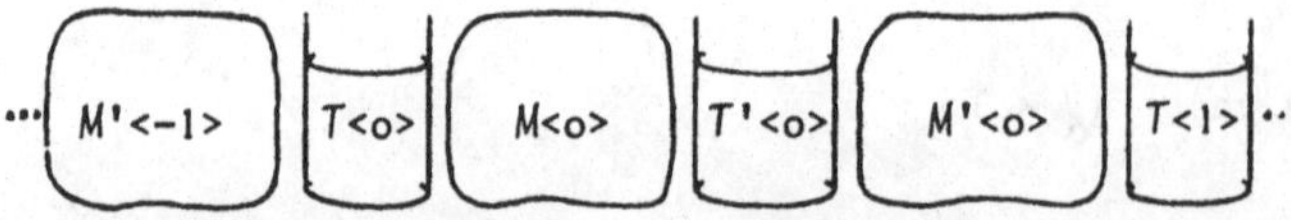

exhaust F(λ,p,q)-mod, and they have the following separation property: any of these classes separates the additive subcategory given by the module classes to its left from the additive subcategory given by the module classes to its right. Note that the modules in M<n> live on Δ<n> ∪ Δ'<n>, those in M'<n> live on Δ'<n> ∪ Δ<n+1>, the modules T<n> live on Δ'<n-1> ∪ Δ<n> ∪ Δ'<n>, those in T'<n> live on Δ<n> ∪ Δ'<n> ∪ Δ<n+1>. Altogether, we see that the indecomposable F(λ,p,q)-modules have bounded support.

The structure of the module classes M<n> and M'<n> is known only in particular cases. Consider the star $\mathbb{T} = \mathbb{T}_{p_0+q_0-1,\dots,p_t+q_t-1}$. In case $\mathbb{T}$ is one of the Dynkin graphs $\mathbb{A}_m$, $\mathbb{D}_m$, $\mathbb{E}_6$, $\mathbb{E}_7$ or $\mathbb{E}_8$, then any M<n> and M'<n> is a single component, and it is of the form $\mathbb{Z}\tilde{\mathbb{T}}$ (in case $\mathbb{T} = \mathbb{A}_m$, say $p_i+q_i = 2$ for all $i \geq 2$, so that $\tilde{\mathbb{T}}$ is not a tree, we have to specify which orientation of $\tilde{\mathbb{T}}$ has to be taken: it is just that of $\tilde{\mathbb{A}}_{p_0+q_0-1,p_1+q_1-1}$).Thus, in this case the category F(λ,p,q)-mod has the following shape: there are the tubular families T<n> and T'<n>, and in between there are sort of connecting components of the form $\mathbb{Z}\tilde{\mathbb{T}}$.

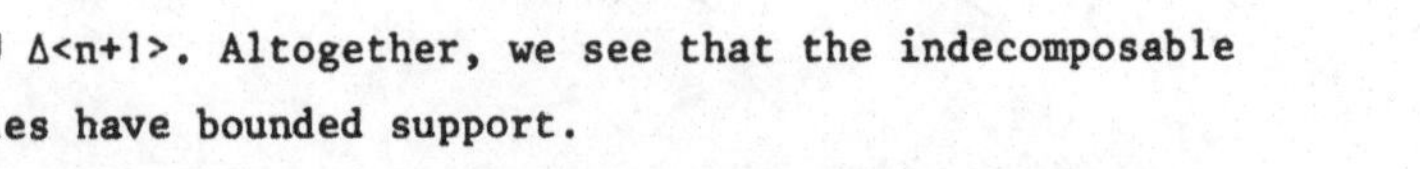

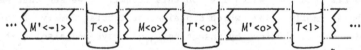

Next consider the case of $\mathbb{T}$ being a Euclidean graph, thus one of $\tilde{\mathbb{D}}_4$, $\tilde{\mathbb{E}}_6$, $\tilde{\mathbb{E}}_7$ and $\tilde{\mathbb{E}}_8$. In this case, all components in M<n> and in M'<n> are again tubes, all even regular, these tubes form again $\mathbb{P}_1k$-families of type $(p_0+q_0-1,\dots,p_t+q_t-1)$ and the set of families in any M<n> and in any M'<n> may be indexed in a rather natural way by the rational numbers q with $o < q < 1$, see [HR].

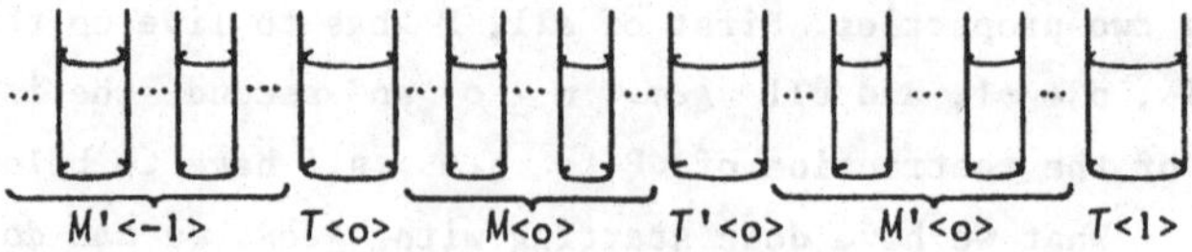

We return now to a general Frobenius category. The category R-mod usually is no longer an abelian category, but it carries some additional structure which seems to be similarly useful: it is the underlying category of a triangulated category. The notion of a triangulated category was introduced by Verdier [V]. The relevant features of R-$\underline{\text{mod}}$ as a triangulated category were observed by Heller [He], however the system of axioms as proposed by Verdier was not yet available; the fact that R-$\underline{\text{mod}}$ actually satisfies all the axioms required by Verdier was noted only recently by Happel [Ha]. We will not repeat all the axioms of a triangulated category and refer directly to Verdier [V]. But let us sketch some of the basic principles of triangulated categories in general, and the way they are used when dealing with R-$\underline{\text{mod}}$. First of all, a triangulated category is an additive category $\mathcal{A}$ together with an automorphism T of $\mathcal{A}$ and a set $\mathcal{T}$ of sextuples of the form (X,Y,Z,u,v,w) where $u : X \longrightarrow Y$, $v : Y \longrightarrow Z$, $w : Z \longrightarrow T(X)$ are maps in $\mathcal{A}$, the elements of $\mathcal{T}$ being called triangles. Of course, the set $\mathcal{T}$ is supposed to be closed under isomorphisms (a map from (X,Y,Z,u,v,w) to (X',Y',Z',u',v',w') is of the form (f,g,h), where $f : X \longrightarrow X'$, $g : Y \longrightarrow Y'$, $h : Z \longrightarrow Z'$ are maps in $\mathcal{A}$ satisfying $ug = fu'$, $vh = gv'$ and $wT(f) = hw'$). In the case of $\mathcal{A}$ = R-$\underline{\text{mod}}$, one takes for T the suspension functor of Heller: for any object X, choose an injective module $I(X)$ with submodule X, and let $T(X) = I(X)/X$. (There are some set-theoretical sutleties in order to ensure that T actually is an automorphism of R-$\underline{\text{mod}}$ as required and not only a self-equivalence). Given a map $\alpha : X \longrightarrow Y$ in R-mod, consider the induced exact sequence

$$\begin{array}{ccccccccc} 0 & \longrightarrow & X & \xrightarrow{\iota} & I(X) & \xrightarrow{\pi} & T(X) & \longrightarrow & 0 \\ & & \downarrow{\scriptstyle\alpha} & & \downarrow{\scriptstyle\bar{\alpha}} & & \| & & \\ 0 & \longrightarrow & Y & \xrightarrow{\beta} & Z & \xrightarrow{\gamma} & T(X) & \longrightarrow & 0 \end{array}$$

where $\iota = \iota_X$ denotes the inclusion, $\pi = \pi_X$ the projection map. Then $(X,Y,Z,\underline{\alpha},\underline{\beta},\underline{\gamma})$ is called a standard triangle, and, by definition $\mathcal{T}$ is the class of sextuples which are isomorphic to standard triangles. For example, with the standard triangle $(X,Y,Z,\underline{\alpha},\underline{\beta},\underline{\gamma})$ above, also the isomorphic one $(X, Y \oplus I(X), Z, \underline{[\alpha\iota]}, \underline{\begin{bmatrix}\beta\\ -\bar{\alpha}\end{bmatrix}}, \underline{\gamma})$ is a triangle (in R-$\underline{\text{mod}}$, there is the isomorphism $(1,\begin{bmatrix}1\\0\end{bmatrix},1)$ from the first to the second sextuple!) Given any exact sequence

$$E = (0 \longrightarrow U \xrightarrow{\mu} V \xrightarrow{\varepsilon} W \longrightarrow 0)$$

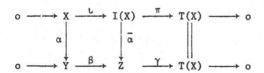

in R-mod, there is $\bar{\mu} : V \longrightarrow I(U)$ with $\mu\bar{\mu} = \iota_U$, and therefore a commutative diagram

$$\begin{array}{ccccccccc}
0 & \longrightarrow & U & \xrightarrow{\mu} & V & \xrightarrow{\varepsilon} & W & \longrightarrow & 0 \\
 & & \| & & \downarrow{\scriptstyle \bar{\mu}} & & \downarrow{\scriptstyle \nu} & & \\
0 & \longrightarrow & U & \xrightarrow{\iota} & I(U) & \xrightarrow{\pi} & T(U) & \longrightarrow & 0 .
\end{array}$$

Note that $E \longmapsto w(E) = -\underline{\nu}$ yields a map $w : \mathrm{Ext}^1(W,U) \longrightarrow \underline{\mathrm{Hom}}(W,T(U))$ which is known to be bijective. On the other hand, we can rearrange these maps to the following commutative diagram with exact rows:

$$\begin{array}{ccccccccc}
0 & \longrightarrow & U & \xrightarrow{\iota} & I(U) & \xrightarrow{\pi} & T(U) & \longrightarrow & 0 \\
 & & \downarrow{\scriptstyle \mu} & & \downarrow{\scriptstyle [0\,1]} & & \| & & \\
0 & \longrightarrow & V & \xrightarrow{[\varepsilon\bar{\mu}]} & W \oplus I(U) & \xrightarrow{\left[\begin{smallmatrix} -\nu \\ \pi \end{smallmatrix}\right]} & T(U) & \longrightarrow & 0
\end{array}$$

which shows that $(U,V,W \oplus I(U),\underline{\mu}, \underline{\varepsilon\bar{\mu}}, \left[\begin{smallmatrix} -\nu \\ \pi \end{smallmatrix}\right])$ is a standard triangle. But this standard triangle is isomorphic to $(U,V,W,\underline{\mu},\underline{\varepsilon},-\underline{\nu}) = (U,V,W,\underline{\mu},\underline{\varepsilon},w(E))$, thus also the latter is a triangle. Let us return to the standard triangle $(X,Y,Z,\underline{\alpha},\underline{\beta},\underline{\gamma})$ considered above, and the isomorphic triangle $(X,Y \oplus I(X),Z,\underline{[\alpha\iota]},\underline{\left[\begin{smallmatrix} \beta \\ -\bar{\alpha} \end{smallmatrix}\right]},\underline{\gamma})$. We observe that

$$E_o := (0 \longrightarrow X \xrightarrow{[\alpha\iota]} Y \oplus I(X) \xrightarrow{\left[\begin{smallmatrix} \beta \\ -\bar{\alpha} \end{smallmatrix}\right]} Z \longrightarrow 0)$$

is an exact sequence, and the commutative diagram

$$\begin{array}{ccccccccc}
0 & \longrightarrow & X & \xrightarrow{[\alpha\iota]} & Y \oplus I(X) & \xrightarrow{\left[\begin{smallmatrix} \beta \\ -\bar{\alpha} \end{smallmatrix}\right]} & Z & \longrightarrow & 0 \\
 & & \| & & \downarrow{\scriptstyle \left[\begin{smallmatrix} 0 \\ 1 \end{smallmatrix}\right]} & & \downarrow{\scriptstyle -\gamma} & & \\
0 & \longrightarrow & X & \xrightarrow{\iota} & I(X) & \xrightarrow{\pi} & T(X) & \longrightarrow & 0
\end{array}$$

shows that $w(E_o) = \underline{\gamma}$. This demonstrates that any triangle in R-$\underline{\mathrm{mod}}$ is isomorphic to a triangle of the form $(U,V,W,\underline{\mu},\underline{\varepsilon},w(E))$, where

$$E = (0 \longrightarrow U \xrightarrow{\mu} V \xrightarrow{\varepsilon} W \longrightarrow 0)$$

is an exact sequence.

By the construction above, we see that any map u in R-mod occurs as a first map in some triangle (X,Y,Z,u,v,w), and in case u is an identity map, then Z is zero (in R-mod). this is one of the conditions required in a triangulated category. Note that Z is determined by u up to an isomorphism, but given two triangles (X,Y,Z,u,v,w) and (X,Y,Z',u,v',w'), there may be different maps $h_1,h_2 : Z \longrightarrow Z'$ in R-mod with $(1,1,h_1)$, $(1,1,h_2)$ both being isomorphisms of triangles. The non-unicity of maps when working inside a triangulated category is one of the phenomena, which makes such a category completely different to an abelian category! Given two triangles (X,Y,Z,u,v,w), (X',Y',Z',u',v',w') and $f : X \longrightarrow X'$, $g : Y \longrightarrow Y'$ with $ug = fu'$, then one of the axioms of a triangulated category requires the existence of some $h : Z \longrightarrow Z'$ (again not necessarily a unique one) such that (f,g,h) is a map of triangles. There are two other axioms which have to be checked. One of these, the socalled octahedral axiom, starts with two composable maps, say $u_1 : X_1 \longrightarrow X_2$, $u_2 : X_2 \longrightarrow X_3$, and triangles which have u_1,u_2, and u_1u_2 as first map, respectively, and shows the relation between these triangles. The other one is the most surprising: it provides a rotation-symmetry: given a triangle (X,Y,Z,u,v,w), then also $(Y,Z,T(X),v,w,-T(u))$ is a triangle (there is also the converse, but this follows from the axioms). This shows that the properties mentioned with respect to the first map of a triangle have corresponding counterparts for the remaining maps. On the other hand, we also see that the properties mentioned with respect to the second or the third map of a triangle hold true also for the first map. In particular, given a triangle (X,Y,Z,u,v,w), up to isomorphism, we may represent u by a monomorphism, or an epimorphism of R-mod, or we may consider it as an element in $\mathrm{Ext}^1(X,T^{-1}Y)$.

In any triangulated category A, we may speak of <u>Auslander-Reiten triangles</u>, these are triangles (X,Y,Z,u,v,w) with both X, and Z indecomposable, $w \neq o$, and the following equivalent conditions are satisfied: (i) for all $f : X \longrightarrow V$, f not split mono, there exists $f' : Y \longrightarrow V$ with $uf' = f$; (ii) for all $g : W \longrightarrow Z$, g not split epi, there exists $g' : W \longrightarrow Y$ with $g'v = g$; (iii) if $h_1 : U_1 \longrightarrow Z$, h_1 not split epi, then $h_1w = o$; (iv) if $h_2 : TX \longrightarrow U_2$, h_2 not split mono, then $wh_2 = o$. In case (X,Y,Z,u,v,w) is an Auslander-Reiten triangle, the objects X and Z determine each other (up to isomorphism), and we write $X = \tau Z$, $Z = \tau^- X$, and τ is called the Auslander-Reiten translation; also, if we decompose $Y = \oplus Y_i$ with indecomposable objects Y_i, then the induced morphisms $X \longrightarrow Y_i$ are irreducible, and we obtain in this way sufficiently many irreducible morphisms starting in X; similarly, the induced morphisms $Y_i \longrightarrow Z$ are irreducible, and we obtain sufficiently many irreducible morphisms ending in Z. We say that $\mathcal{A}$ has Auslander-Reiten triangles, provided for any indecomposable object of $\mathcal{A}$ there is an Auslander-Reiten triangle where it occurs in the first position, and one where it occurs in the third position. For a triangulated category $\mathcal{A}$ with Auslander-Reiten triangles, we can introduce its Auslander-Reiten quiver in the same way as for module categories, note however that here we obtain a regular translation quiver!

Let us display the Auslander-Reiten triangles in R-mod, where R is a Frobenius algebra. Given any indecomposable non-injective R-module X, let $E = (o \longrightarrow X \xrightarrow{f} Y \xrightarrow{g} Z \longrightarrow o)$ be an Auslander-Reiten sequence starting in X. Then $(X,Y,Z,f,g,w(E))$ is an Auslander-Reiten triangle in R-mod, and any Auslander-Reiten triangle is isomorphic to such a triangle. In particular, we see that R-mod has Auslander-Reiten triangles. In case $R = \hat{A}$, where A is a finite-dimensional algebra, some of the Auslander-Reiten sequences in A-mod are still Auslander-Reiten sequences in $\hat{A}$-mod, thus give rise to an Auslander-Reiten triangle in $\hat{A}$-mod:

<u>Lemma</u>. Let A be a finite-dimensional algebra, and $E := (o \longrightarrow X \xrightarrow{f} Y \xrightarrow{g} Z \longrightarrow o)$ an Auslander-Reiten sequence in A-mod. Then the following conditions are equivalent:

(i) E is an Auslander-Reiten sequence in $\hat{A}$-mod.

(ii) inj.dim.$X = 1$, proj.dim.$Z = 1$.

(iii) $\mathrm{Hom}_A(I,X) = o$ for any injective A-module I, and $\mathrm{Hom}_A(Z,P) = o$ for any projective A-module P.

Auslander-Reiten sequences with these equivalent properties may be said to be <u>conservative</u>. The equivalence of the conditions (ii) and (iii) is quite well-known. For the equivalence of (i) and (iii), one may use a generalized version of [Ri2], 2.5.5. We note that for a conservative Auslander-Reiten sequence $E = (o \longrightarrow X \longrightarrow Y \longrightarrow Z \longrightarrow o)$ in A-mod, we have $\underline{\dim} X = -(\underline{\dim}\, Z)C_A^{-T}C_A$, provided C_A is invertible.

(The transformation $\Phi_A = -C_A^{-T}C_A$ which yields $\underline{\dim}\, X = (\underline{\dim}\, Z)\Phi_A$ is usually called the Coxeter transformation for A).

In particular, we see that the Auslander-Reiten sequences inside a sincere separating regular tubular family T of A-mod remain Auslander-Reiten sequences in $\hat{A}$-mod. Thus T remains a set of components of $\hat{A}$-mod. Actually, T also remains to be separating, but, of course, not sincere. With T also the module classes of $\hat{A}$-mod obtained from T by applying powers of the shift ν are separating regular tubular families, thus in this case $\hat{A}$-mod has countably many separating regular tubular families.

Similarly, assume that there is a slice S in A-mod, let ${}_AS$ be a slice module in S, and $k\Delta = \mathrm{End}({}_AS)$. As a module class in $\hat{A}$-mod, S still will be path closed and will satisfy the following property: given any Auslander-Reiten sequence $o \longrightarrow X \longrightarrow Y \longrightarrow Z \longrightarrow o$ in $\hat{A}$-mod, at most one of X,Z belongs to S, and one of X,Z belongs to S in case an indecomposable direct summand of Y is in S. It follows that S belongs to one component Γ of $\hat{A}$-mod, and that the corresponding stable translation quiver Γ_s is of the form $\Gamma_s = \mathbb{Z}\Delta$.

We have seen above that for any Frobenius algebra R, the stable category R-$\underline{\mathrm{mod}}$ can be made into a triangulated category. There is a more general result which includes this as a special case: Instead of dealing with the abelian category R-mod (and assuming that the projective objects and the injective objects coincide), one may start with an arbitrary exact category (A,S). (Here, A is an additive category, which is embedded into some abelian category A' as a full and extension-closed subcategory, and S is the set of all sequences $o \longrightarrow X \xrightarrow{u} Y \xrightarrow{e} Z \longrightarrow o$ in A which are exact when considered as sequences in A'; given such an element of S, the map u is called a proper mono, the map e a proper epi.) Given an exact category (A,S), we say that P is S-projective provided P has the usual lifing property with respect to all proper epis, and we say that Q is S-injective provided Q has the usual extension property with respect to all proper monos. We call S a <u>Frobenius structure</u> on A provided (A,S) is an exact category, the S-projective objects in A coincide with the S-injective objects, and there are enough S-projective and enough S-injective objects (this means that for any object X in A, there is a proper epi $P \longrightarrow X$ with P S-projective, and a proper mono $X \longrightarrow Q$ with Q S-injective). The corresponding stable category $\underline{A}$ has the same objects as A, and $\underline{\mathrm{Hom}}_A(X,Y) = \mathrm{Hom}_A(X,Y)/\sim$, with $f \sim g$ iff $f-g$ factors through an S-projective object. There is the following general result [Ha]: if S is a Frobenius structure on A, then $\underline{A}$ is the underlying category of a triangulated category.

A special case of this result is the following one: We start with any additive category A. A sequence of the form $o \longrightarrow X \xrightarrow{[1\,o]} X \oplus Z \xrightarrow{\left[\begin{smallmatrix} o \\ 1 \end{smallmatrix}\right]} Z \longrightarrow o$, and isomorphic ones, are said to be split exact. Denote by $C^b(A)$ the category of all bounded complexes over A (a bounded complex is of the form $X^\cdot = (X^i, d^i)_i$ where X^i are objects of A, indexed over $\mathbb{Z}$, all but only finitely many being non-zero, and $d^i_{X^\cdot} = d^i : X^i \longrightarrow X^{i+1}$ are maps in A satisfying $d^i d^{i+1} = o$ for all $i \in \mathbb{Z}$. Let S be the set of all sequences $o \longrightarrow X^\cdot \xrightarrow{f^\cdot} Y^\cdot \xrightarrow{g^\cdot} Z^\cdot \longrightarrow o$ in $C^b(A)$ such that the sequences $o \longrightarrow X^i \xrightarrow{f^i} Y^i \xrightarrow{g^i} Z^i \longrightarrow o$ are split exact. Then S is a Frobenius structure on $C^b(A)$, and the usual homotopy category $K^b(A)$ of $C^b(A)$ is just the stable category $\underline{C^b(A)}$ of $C^b(A)$ with respect to S. The automorphism T required in the definition of a triangulated category is given by the shift T wich sends the complex $X^\cdot = (X^i, d^i)_i$ to $T(X^\cdot)$ where $T(X^\cdot)^i = X^{i+1}$ and $d^i_{T(X^\cdot)} = -d^{i+1}_{X^\cdot}$. Thus, the well-known fact that the homotopy category $K^b(A)$ of bounded complexes over A can be made into a triangulated category is a special case of the general result concerning Frobenius structures.

In particular, we are interested in $K^b(A\text{-proj})$, where A is a finite-dimensional algebra (and A-proj the full subcategory of A-mod of all projective modules). In case A has finite global dimension, we may define the <u>derived category</u> $D^b(A)$ to be just $K^b(A\text{-proj})$. More precisely, $D^b(A)$ is usually defined as the triangulated category obtained from $K^b(A\text{-mod})$ by formally inverting all maps which induce an isomorphism in cohomology, thus there is a functor $\Phi : K^b(A\text{-mod}) \longrightarrow D^b(A)$ with the following properties: first of all, if $X^\cdot, Y^\cdot$ are bounded complexes, $f^\cdot : X^\cdot \longrightarrow Y^\cdot$ a map with $H^i(f^\cdot)$ an isomorphism for all i, then $\Phi(\underline{f^\cdot})$ of the homotopy class $\underline{f^\cdot}$ of $f^\cdot$ under Φ is invertible in $D^b(A)$. Second, any other functor $\Phi' : K^b(A\text{-mod}) \longrightarrow \mathcal{D}$ having this property can be factored through Φ. Now, $K^b(A\text{-proj})$ is a full subcategory of

K^b(A-mod),and, in case A has finite global dimension, the composition of functors

$$K^b(A\text{-proj}) \hookrightarrow K^b(A\text{-mod}) \xrightarrow{\Phi} D^b(A)$$

is an equivalence of triangulated categors. Similarly, for A of finite global dimension, also the composition of functors

$$K^b(A\text{-inj}) \hookrightarrow K^b(A\text{-mod}) \xrightarrow{\Phi} D^b(A)$$

is an equivalence of categories (here, A-inj is the full subcategory of A-mod of all injective modules). Note that there are obvious embeddings of A-mod into K^b(A-mod), and using then Φ, into $D^b(A)$ which we may denote by $-[n]$, where $n \in \mathbb{Z}$. Given an A-module M, let $M[n]$ be the complex with $M[n]^i = M$ for $i = -n$, and $= o$ otherwise, or better, the corresponding image under Φ in $D^b(A)$. We will identify A-mod with the image of the functor $-[o]$. Note that if

$$o \longrightarrow M \longrightarrow I^o \longrightarrow I^1 \longrightarrow \dots \longrightarrow I^r \longrightarrow o$$

is an injective resolution of M, then $M = M[o]$ is isomorphic in $D^b(A)$ to the image of the complex

$$\dots \longrightarrow o \longrightarrow I^o \longrightarrow I^1 \longrightarrow \dots \longrightarrow I^r \longrightarrow o \rightarrow \dots$$

under Φ. Similarly, if

$$o \longrightarrow P^{-r} \longrightarrow \dots \longrightarrow P^{-1} \longrightarrow P^o \longrightarrow M \longrightarrow o$$

is a projective resolution of M, then $M = M[o]$ is isomorphic in $D^b(A)$ to the image of the complex

$$\dots \longrightarrow o \longrightarrow P^{-r} \longrightarrow \dots \longrightarrow P^{-1} \longrightarrow P^o \longrightarrow o \rightarrow \dots$$

under Φ. Of course, the different embeddings $-[n]$ of A-mod into $D^b(A)$ are related by the powers of T operating on $D^b(A)$, we have $T(M[n]) = M[n+1]$. For given A-modules M,N, we may identify $\mathrm{Hom}_{D^b(A)}(M[m],N[n])$ with $\mathrm{Ext}_A^{n-m}(M,N)$ (again, $\mathrm{Ext}^o = \mathrm{Hom}$, and $\mathrm{Ext}^i = o$ for $i < o$).

Let us exhibit the Auslander-Reiten triangles in D^b(A-mod). We denote by $D = \mathrm{Hom}_k(-,k)$ the duality with respect to the base field k.

The endofunctor $\nu = D\,\mathrm{Hom}_A(-,{}_AA)$ of A-mod is called the Nakayama functor, it defines an equivalence from A-proj to A-inj. For $X,Y \in$ A-mod, there is a natural map $\alpha_{XY} : D\,\mathrm{Hom}(X,Y) \longrightarrow \mathrm{Hom}(Y,\nu X)$ which is the composition $\alpha_{XY} = \alpha'_{XY}\alpha''_{XY}$, where α'_{XY} is the dual of the map $\mathrm{Hom}(X,{}_AA) \otimes Y \longrightarrow \mathrm{Hom}(X,Y)$ which sends $f \otimes y$ to $(x \longmapsto (xf)y)$, for $f : X \longrightarrow {}_AA$, $y \in Y$, $x \in X$, and where α''_{XY} is the adjunction map. Note that for X a projective A-module, the map α'_{XY}, and therefore α_{XY} itself, is bijective. We extend ν to an endofunctor of C^b(A-mod), and of K^b(A-mod), and we see that its restriction to K^b(A-proj) provides an equivalence $\nu : K^b(\text{A-proj}) \longrightarrow K^b(\text{A-inj})$. Also, given $X^\cdot, Y^\cdot \in C^b$(A-mod), there is the corresponding natural map $\alpha_{X^\cdot Y^\cdot} : D\,\mathrm{Hom}(X^\cdot,Y^\cdot) \to \mathrm{Hom}(Y^\cdot,\nu X^\cdot)$. Assume now that A has finite global dimension, so that any object in $D^b(A)$ can be written in the form $P^\cdot$, where $P^\cdot$ is a bounded complex of projective A-modules. Assume that $P^\cdot$ is indecomposable, and let $\varphi \in D\,\mathrm{Hom}(P^\cdot,P^\cdot)$ be a non-zero linear form on $\mathrm{Hom}(P^\cdot,P^\cdot) = \mathrm{End}(P^\cdot)$ which vanishes on $\mathrm{rad}\,\mathrm{End}(P^\cdot)$. Consider $\alpha_{P^\cdot P^\cdot}(\varphi)$, this is a non-zero map $P^\cdot \longrightarrow \nu P^\cdot$ which has the following properties: if $X^\cdot$ is an indecomposable object of $D^b(A)$ and $\xi : X^\cdot \longrightarrow P^\cdot$ is non-invertible, or $\eta : \nu P^\cdot \longrightarrow X^\cdot$ is non-invertible, then $\xi\alpha_{P^\cdot P^\cdot}(\varphi) = 0$, or $\alpha_{P^\cdot P^\cdot}(\varphi)\eta = 0$, respectively. Take a triangle in $D^b(A)$ whose third map is $\alpha_{P^\cdot P^\cdot}(\varphi)$, say $(T^{-1}\nu P^\cdot, Y, P^\cdot, u, v, \alpha_{P^\cdot P^\cdot}(\varphi))$, then this is an Auslander-Reiten triangle, and, up to isomorphism, one obtains all Auslander-Reiten triangles in $D^b(A)$ in this way. Note that we see that the Auslander-Reiten translation on K^b(A-proj) ($\approx D^b(A)$) is given by the functor $\tau := T^{-1}\nu$.

Following Happel [Ha], we consider one example in detail. Let A be a hereditary algebra. In this case (and only in this case), the images of the various full embeddings $-[n] : \text{A-mod} \to D^b(A)$ exhaust the category $D^b(A)$: any indecomposable object of $D^b(A)$ belongs to one (and only one) such subcategory. This is of course straightforward to see since obviously any complex in C^b(A-proj) can be written as a direct sum of complexes of the form $\ldots 0 \longrightarrow P^{n-1} \xrightarrow{d} P^n \longrightarrow 0 \ldots$ with d a monomorphism, and in $D^b(A)$ this complex is isomorphic to the complex $(\mathrm{Cok}\,d)[-n]$. Note that (for A hereditary!) the embedding $-[n]$ preserves irreducibility of maps and that any Auslander-Reiten sequence $0 \longrightarrow X \xrightarrow{f} Y \xrightarrow{g} Z \longrightarrow 0$ in A-mod gives rise to Auslander-Reiten triangles $(X[n],Y[n],Z[n],f[n],g[n],h_n)$ where $h_n : Z[n] \to X[n+1]$.

Consequently, we may visualize $D^b(A)$ as follows: Take copies $\Gamma_A[n]$ of Γ_A, indexed by $n \in \mathbb{Z}$, and connect these copies together using additional arrows and extending the translation τ to all vertices. In fact, each arrow $a \xrightarrow{\alpha} b$ in Δ gives rise to an irreducible morphism $Q(a)[n] \longrightarrow P(b)[n+1]$, in the following way: instead of writing down a map $Q(a)[n] \longrightarrow P(b)[n+1]$, we may as well exhibit an element in $\mathrm{Ext}^1(Q(a),P(b))$. Let $V(\alpha)$ be the representation of Δ which first of all coincides on the support of $P(b)$ with $P(b)$, on the support of $Q(a)$ with $Q(a)$, which second has $V(\alpha)_\alpha = 1_k$, and finally, which elsewhere is given by zero spaces and zero maps. Then there is a non-split exact sequence

$$o \longrightarrow P(b) \longrightarrow V(\alpha) \longrightarrow Q(a) \longrightarrow o$$

and this, in fact, corresponds to an irreducible morphism $Q(a)[n] \longrightarrow P(b)[n+1]$, for every $n \in \mathbb{Z}$. Also, for every $n \in \mathbb{Z}$, and any vertex c of Δ, there is a triangle of the form $(Q(c)[n],Y,P(c)[n+1],u_n,v_n,w[n+1])$, where $w \in \mathrm{Hom}_A(P(c),Q(c))$ is a non-zero map, and one easily checks that this indeed is an Auslander-Reiten triangle, and that $Y = (\bigoplus_{a\to c} Q(a)) \oplus (\bigoplus_{c\to b} P(b))$. This shows that the additional arrows which we need in order to connect the various $\Gamma_A[n]$, are of the form $[Q(a)[n]] \longrightarrow [P(b)[n+1]]$, one for each arrow $a \longrightarrow b$, and that $\tau[P(c)[n+1]] = [Q(c)[n]]$. In case Δ is of Dynkin type, so that Γ_A is finite, the various copies $\Gamma_A[n]$ all are connected together, and we obtain just $\mathbb{Z}\Delta^*$ (and, actually, $D^b(A)$ is equivalent, as a category, to $k(\mathbb{Z}\Delta^*)$). For example, for the following quiver Δ of type $\mathbb{E}_6$, and $A = k\Delta^*$, the copies $\Gamma_A[-1]$, $\Gamma_A[o]$ and $\Gamma_A[1]$ are connected as indicated:

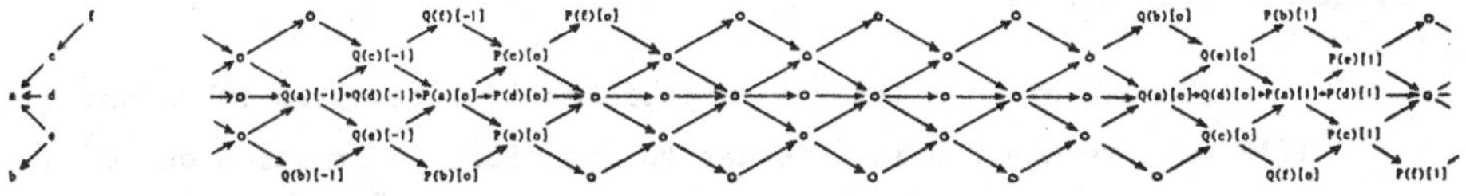

Δ $\Gamma_A[-1]$ $\Gamma_A[o]$ $\Gamma_A[1]$

In case Δ is connected and not of Dynkin type, the preinjective component of $\Gamma_A[n-1]$ is connected with the preprojective component $\Gamma_A[n]$ and together they form a component of the form $\mathbb{Z}\Delta^*$. We denote by $C[n]$ the full subcategory whose indecomposable objects are those in the preinjective component of $\Gamma_A[n-1]$ and those in the preprojective component of $\Gamma_A[n]$, (then $C[n]$ is actually equivalent, as a category, to $k(\mathbb{Z}\Delta^*)$), and by $R[n]$ the image of the full subcategory of all regular A-modules under $-[n]$. With these notations, $D^b(A)$ can be visualised as follows:

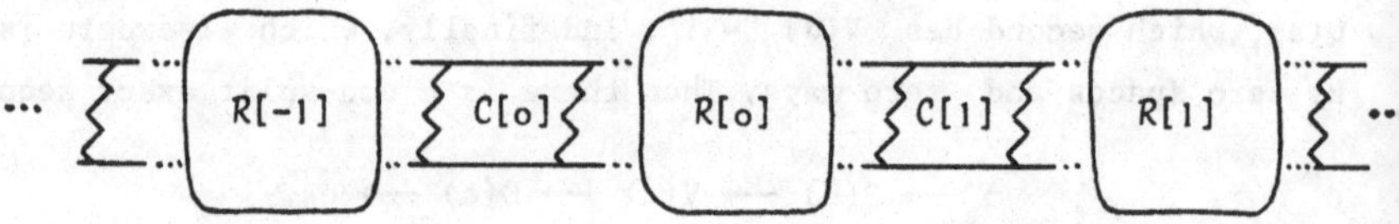

Note that if X belongs to $C[n]$, and Y is indecomposable with $\mathrm{Hom}(X,Y) \neq o$, then Y belongs to $C[n]$ or $R[n]$ or $C[n+1]$. Similarly, if X belongs to $R[n]$ and Y is indecomposable with $\mathrm{Hom}(X,Y) \neq o$, then Y belongs to $R[n]$ or $C[n+1]$ or $R[n+1]$.

Given any finite dimensional algebra A, there are two triangulated categories obtained from A-mod, namely $\hat{A}$-<u>mod</u> and $D^b(A)$. In case A has finite global dimension, there is the following theorem of Happel which asserts that these categories are equivalent, not only as categories, but even as triangulated categories. [Given two triangulated categories A, B, an equivalence of triangulated categories is given by an equivalence $F : A \longrightarrow B$ of categories and an equivalence $\eta : FT \longrightarrow TF$ of functors such that for any triangle (X,Y,Z,u,v,w) in A, the sextuple $(FX,FY,FZ,Fu,Fv,Fw\cdot\eta_X)$ is a triangle in B].

<u>Theorem</u> (Happel). Assume A is a finite dimensional algebra of finite global dimension. Then $\hat{A}$-<u>mod</u> and $D^b(A)$ are equivalent as triangulated categories.

The use of this theorem is twofold: if interested in $\hat{A}$-mod or in T(A)-mod, we may use it in order to transform information on $D^b(A)$ to $\hat{A}$-<u>mod</u>, thus the report above on the structure of $D^b(A)$ for A being hereditary yields the same information for $\hat{A}$-mod. Similarly, if we are interested in $D^b(A)$, we may work instead with the categories $\hat{A}$-mod and $\hat{A}$-<u>mod</u>; for example, for A a canonical algebra, our investigation of $\hat{A}$-mod yields a clear description of $D^b(A)$ in this case. Note that one advantage of considering $\hat{A}$-<u>mod</u> instead of $D^b(A)$ lies in the fact that

$\underline{\hat{A}\text{-mod}}$ is obtained from the category $\hat{A}$-mod by a rather easy modification, and $\hat{A}$-mod is abelian!

We should add the following warning: there are given countably many embeddings of A-mod both into $\underline{\hat{A}\text{-mod}}$ (denoted by $-\langle n\rangle$, $n \in \mathbb{Z}$) and into $D^b(A)$ (denoted by $-[n]$, $n \in \mathbb{Z}$, and Happel's equivalence $D^b(A) \longrightarrow \underline{\hat{A}\text{-mod}}$ is constructed by first identifying the subcategory (A-mod)[o] of $D^b(A)$ with the subcategory (A-mod)$\langle o\rangle$ of $\underline{\hat{A}\text{-mod}}$ and extending this to an equivalence $D^b(A)$ and $\underline{\hat{A}\text{-mod}}$. We should warn that the remaining full subcategories (A-mod)[n] and (A-mod)$\langle n\rangle$, $n \neq o$ will <u>not</u> coincide, in general. In fact, the two shift functors ν and T (which are both defined on $\hat{A}$-mod) are related by the Auslander-Reiten translation τ: we have $\nu = T^2\tau$.

A special case of Happel's theorem has been known before and is of particular interest. Investigations by Beilinson [Bei] and Bernstein-Gelfand-Gelfand [BGG] have related the category $V = V_n$ of vector bundles over the projective space $\mathbb{P}_n k$ to problems in linear algebra. Both papers provide a description of the derived category $D^b(\mathrm{Coh}\ \mathbb{P}_n k)$ of bounded complexes of coherent sheaves over $\mathbb{P}_n k$, and the theorem above yields a direct interrelation between these two descriptions. Consider the following quiver $\tilde{\Delta}$

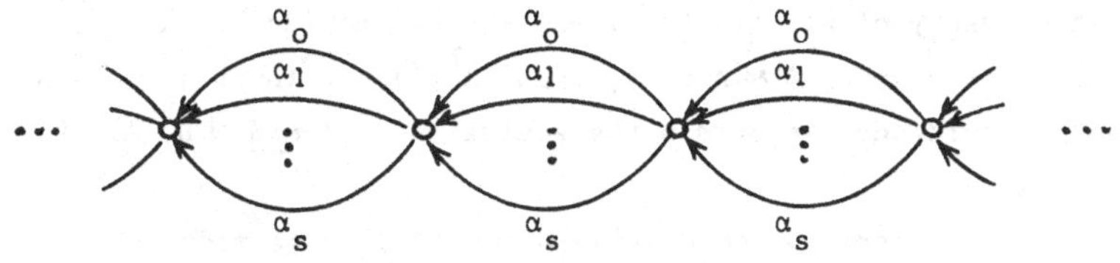

with set of vertices the integers $\mathbb{Z}$; for any vertex $a \in \mathbb{Z}$, there are $n+1$ arrows $\alpha_s = \alpha_s^{(a)} : a \longrightarrow a+1$, $o \leq s \leq n$. Denote $\tilde{\Lambda}$ the opposite of the path algebra of $\tilde{\Delta}$ over the field k modulo the relations

$$\alpha_s\alpha_s = o, \quad \alpha_s\alpha_t+\alpha_t\alpha_s = o, \quad \text{for all } s,t \text{ with } s < t.$$

Note that $\tilde{\Lambda}$-mod is just the category Λ-grmod of graded Λ-modules, where Λ is the exterior algebra on the vectorspace $\Lambda_1 = k^{n+1}$, with the usual grading, thus $\Lambda_i = \Lambda^i(k^{n+1})$ [one may observe that $\tilde{\Lambda}$ is a Galois covering of Λ; on $\tilde{\Lambda}$ there is an obvious $\mathbb{Z}$-action by shifting the quiver, and $\Lambda = \tilde{\Lambda}/\mathbb{Z}$]. It is easy to see that $\tilde{\Lambda}$ is a Frobenius algebra. Indeed, we have $P_{\tilde{\Lambda}}(a) = Q_{\tilde{\Lambda}}(a-n-1)$, for any $a \in \mathbb{Z}$.

Given $a \leq b$ in $\mathbb{Z}$, denote by $\tilde{\Lambda}_{ab}$ the restriction of $\tilde{\Lambda}$ to the full subquiver of $\tilde{\Delta}$ with vertices x satisfying $a \leq x \leq b$, and let $A = \tilde{\Lambda}_{on}$. We claim that $\hat{A} = \tilde{\Lambda}$. Note that A has a unique simple projective module, namely $P_A(o)$, and we may identify the one-point extension $A[Q_A(o)]$ with $\tilde{\Lambda}_{o,n+1}$. Inductively, we see that $\tilde{\Lambda}_{ob}$, for $b > n$, is obtained from $\tilde{\Lambda}_{o,b-1}$ as one-point extension, using the indecomposable injective $\tilde{\Lambda}_{o,b-1}$-module with socle at $b-n-1$. The dual process of forming successively one-point coextensions finally shows that $\hat{A} = \tilde{\Lambda}$. The category

$$\Lambda\text{-grmod} = \tilde{\Lambda}\text{-mod} = \hat{A}\text{-mod}$$

is used by Bernstein-Gelfand-Gelfand in order to describe $D^b(\text{Coh}\,\mathbb{P}_n k)$ they construct an equivalence

$$\hat{A}\text{-}\underline{\text{mod}} \approx D^b(\text{Coh}\,\mathbb{P}_n k).$$

On the other hand, the description of $D^b(\text{Coh}\,\mathbb{P}_n k)$ given by Beilinson is

$$D^b(\text{Coh}\,\mathbb{P}_n k) \approx D^b(A).$$

[The actual statement of Beilinson is $D^b(\text{Coh}\,\mathbb{P}_n k) \approx K^b(\mathcal{A})$, where $\mathcal{A}$ is the additive category of all finite direct sums of copies $P_{\tilde{\Lambda}}(a)$, $o \leq a \leq n$, but, of course, $\mathcal{A} \approx A\text{-proj}$, thus $K^b(\mathcal{A}) \approx K^b(A\text{-proj}) \approx D^b(A)$.] Combining both assertions, we obtain the equivalence $\hat{A}\text{-}\underline{\text{mod}} \approx D^b(A)$ for this special A.

We return now to the consideration of tilting modules.

<u>Theorem</u> (Happel). Let A be a finite-dimensional algebra of finite global dimension, and ${}_AT$ a tilting module, with $B = \text{End}({}_AT)$. Then $D^b(A)$ and $D^b(B)$ are equivalent as triangulated categories.

An equivalence $D^b(A) \longrightarrow D^b(B)$ is obtained as follows: the functor $F = \text{Hom}_A({}_AT_B, -) : A\text{-mod} \longrightarrow B\text{-mod}$ is left exact, thus there is the right derived functor $\underline{\underline{R}}F : D^b(A) \longrightarrow D^b(B)$ [on the full and dense subcategory $K^b(A\text{-inj})$ of $D^b(A)$, the functor $\underline{\underline{R}}F$ is defined as follows: $I^{\cdot} \in K^b(A\text{-inj})$ is sent to the complex $(\underline{\underline{R}}F)(I^{\cdot})$ with $((\underline{\underline{R}}F)(I^{\cdot}))^i = FI^i$], and $\underline{\underline{R}}F$ is an equivalence of triangulated categories.

Actually, Happel [Ha] shows that one may consider instead

of tilting modules more generally modules ${}_AT$ of finite projective dimension such that $\mathrm{Ext}^i({}_AT,{}_AT) = o$ for all $i \geq 1$, and such that the T-codimension of ${}_AA$ is finite, and $B = \mathrm{End}({}_AT)$. Note that Miyashita [M] recently has shown that under these assumptions also B is of finite global dimension and that T_B satisfies the corresponding conditions (as a right B-module). [In by-passing, we should stress that the investigations of Miyashita are not restricted to finite-dimensional algebras, but to arbitrary rings (with 1). In this way, the generalized tilting theory established by him also incorporates (as case $r = o$) the Morita equivalence for arbitrary rings].

Combining the two theorems of Happel, we know the structure of $D^b(A)$ for quite a number of different algebras. First of all, if A is a tilted algebra, say $A = \mathrm{End}({}_{k\Delta}T)$, where Δ is a finite quiver without oriented cycles, and ${}_{k\Delta}T$ a tilting module, then $D^b(A) \approx D^b(k\Delta)$, and we have outlined above the structure of $D^b(k\Delta)$. If A is the endomorphism ring of a tilting module over a canonical algebra B, then $D^b(A) \approx D^b(B)$, and again we know much about the structure of $D^b(B)$, this time using the equivalence $D^b(B) \approx \hat{B}$-mod. In case $C = C(\lambda,p)$ is a canonical algebra, where $p = (p_o,\ldots,p_t)$ gives rise to a Dynkin graph $\Delta = \mathbb{T}_{p_o,\ldots,p_t}$, then our description above of $D^b(C)$ coincides with the description of $D^b(k\tilde{\Delta})$. This is now no longer surprising, since one easily shows that C is, in fact, a tilted algebra, which has a slice module with endomorphism ring $k\tilde{\Delta}$.

Let us demonstrate in which way the concepts presented in this lecture can be used in order to obtain insight into tilted algebras. We want to outline the proof of a result announced in the second lecture: that an algebra with slices in two different components has to be concealed. Thus, let A be a (necessarily connected) algebra with slices S and S', which belong to different components of A-mod. Let ${}_AS$ be a slice module for S and $k\Delta = \mathrm{End}({}_AS)$. Thus, $D^b(A)$ and $D^b(k\Delta)$ are equivalent categories and we may use some fixed identification. Also, we may identify $D^b(A)$ and $\hat{A}$-mod, identifying A-mod both with A-mod[o] $\subseteq D^b(A)$ and with A-mod<o> $\subseteq \hat{A}$-mod. We know that S belongs to some component of $\hat{A}$-mod of the form $\mathbb{Z}\Delta$, thus without loss of generality to C[o]. Similarly, S' belongs to some C[m], and since S and S' belong to different components of A-mod, one easily sees that we must have $m \neq o$. Now, S is a separating subcategory of A-mod, say separating P from Q. Let S_1', be an indecomposable module in S'. Assume S_1' belongs to Q, then clearly $\mathrm{Hom}(S,S_1') \neq o$. [For, take an indecomposable projective A-module P with $\mathrm{Hom}(P,S_1') \neq o$, then P in P or S, and factor some non-trivial map from P to S_1' through S]. In this case we must have $m = 1$. Similarly, if S_1' belongs to P, then $m = -1$. Without loss of generality, we assume $m = 1$. Since S is sincere, any indecomposable projective A-module belongs to $C[-1]$ or $R[-1]$ or $C[o]$, since S' is sincere, any indecomposable projective A-module belongs to $C[o]$ or $R[o]$ or $C[1]$, thus the indecomposable projective A-modules all lie in $C[o]$. Now, $C[o]$ is of the from $k(\mathbb{Z}\Delta)$, thus we see that A can be written as the endomorphism ring of a preprojective tilting $k\Delta$-module. This shows that A is a concealed algebra.

References

[Bei] Beilinson, A.: Coherent sheaves on $\mathbb{P}^n$ and problems of linear algebra. Funct. Anal. Appl. 12(1978), 214-216.

[BGG] Bernstein, I.N., Gelfand, I.M., and Gelfand, S.I.: Algebraic bundles over $\mathbb{P}^n$ and problems of linear algebra. Funct. Anal. Appl. 12(1978), 212-214.

[Br] Brenner, S.: A combinatorial characterization of finite Auslander-Reiten quivers. Proceedings Ottawa Conf. 1984. To appear.

[BS] Bautista, R., and Smalø, S.: Nonexistent cycles. Comm. Algebra 11(1983), 1755-1767.

[BSh] Butler, M.C.R., and Shahzamanian, M.: The construction of almost split sequences. III. Modules over two classes of tame local algebras. Math. Ann. 247(1980), 111-122.

[Ha] Happel, D.: On the derived category of a finite dimensional algebra. To appear.

[He] Heller, A.: The loop space functor in homological algebra. Trans. Amer. Math. Soc. 96(1960), 382-394.

[HR] Happel, D., and Ringel, C.M.: The derived category of a tubular algebra. Proceedings Ottawa Conf. 1984. To appear.

[HW] Hughes, D., and Waschbüsch, J.: Trivial extensions of tilted algebras. Proc. London Math. Soc. 46(1983), 347-364.

[M] Miyashita, Y.: Tilting modules of finite projective dimension. To appear.

[Ri1] Ringel, C.M.: Finite dimensional hereditary algebras of wild representation type. Math. Z. 161(1978), 235-255.

[Ri2] Ringel, C.M.: Tame algebras and integral quadratic forms. Springer LNM 1099(1984).

[Ri3] Ringel, C.M.: Components of tilted algebras. To appear.

[RV] Ringel, C.M., and Vossieck, D.: Hammocks. To appear.

[V] Verdier, J.L.: Catégories derivées. In: Cohomologie étale. Springer LNM 569(1977), 262-311.

[W] Webb, P.J.: The Auslander-Reiten quiver of a finite Group. Math. Z. 179(1982), 97-121.

Claus Michael Ringel
Universität Bielefeld
Fakultät für Mathematik
Universitätsstraße
4800 Bielefeld, FRG

A SURVEY OF EXISTENCE THEOREMS FOR ALMOST SPLIT SEQUENCES

Maurice Auslander
Brandeis University
Waltham, MA 02254 U.S.A.

This paper is devoted to describing existence theorems for almost split sequences in various settings, ranging from the category of all modules over an arbitrary ring to locally free coherent sheaves on projective nonsingular curves. We follow closely the first of three lectures we gave at the Durham symposium on the representation theory of algebras. Since these lectures were purely expository, no proofs are given. References to past or future papers where proofs can be found will be given.

We recall that an exact sequence $0 \to A \xrightarrow{f} B \xrightarrow{g} C \to 0$ is called an almost split sequence if:

a) it is not split;

b) if $h : X \to C$ is not a splittable epimorphism, then h can be lifted to B;

c) if $t : A \to Y$ is not a splittable monomorphism, then t can be extended to B.

If $0 \to A \xrightarrow{f} B \xrightarrow{g} C \to 0$ is an almost split sequence, then it is easily seen that End C and End A are local rings which places a serious limitation on the C and A that can be the ends of an almost split sequence. Moreover, the following uniqueness theorem shows that the existence of an almost split sequence $0 \to A \to B \to C \to 0$ establishes a strong connection between A and C.

<u>Theorem 1</u>. Let $0 \to A_i \to B_i \to C_i \to 0$ be two almost split sequences for $i = 1,2$. Then the following statements are equivalent:

i) The sequences are isomorphic;

ii) $A_1 \cong A_2$;

iii) $C_1 \cong C_2$.

So the ends of an almost split sequence determine each other, up to isomorphism.

There are basically two different types of existence theorems for almost split sequences. One type proves their existence without describing how the ends determine each other. The other not only proves the existence of almost split sequences but also gives an independent description of one end in terms of the other. Since a priori knowledge of the ends of an almost split sequence is often helpful in actually finding the sequence, we deal in this talk only with the second type of existence theorem. Basic to describing the ends of an almost split sequence in terms of each other is the notion of the transpose of a finitely presented module. So we begin with this concept.

Let Λ be an arbitrary ring. We denote by $\text{Mod}\,\Lambda$ the category of all Λ-modules and by $\text{mod}\,\Lambda$ the category of finitely presented Λ-modules. Also we denote by $\underline{\text{mod}}\,\Lambda$, the category of finitely presented Λ-modules modulo projectives, the category whose objects are the same as those of $\text{mod}\,\Lambda$ with $\underline{\text{Hom}}_\Lambda(A,B) = \text{Hom}_\Lambda(A,B)/P_\Lambda(A,B)$ where $P_\Lambda(A,B)$ is the subgroup of $\text{Hom}_\Lambda(A,B)$ consisting of those morphisms factoring through finitely generated projective Λ-modules. The transpose is a duality $\text{Tr} : \underline{\text{mod}}\,\Lambda \to \underline{\text{mod}}\,\Lambda^{op}$ which on objects is defined as follows: For each C in $\text{mod}\,\Lambda$ choose a fixed finite projective presentation $P_1 \xrightarrow{f} P_0 \to C \to 0$, with $P_0 \to C$ being the identity if C is projective. Then $\text{Tr}C$ is defined by the exact sequence

$$\text{Hom}_\Lambda(P_0,\Lambda) \xrightarrow{(f,\Lambda)} \text{Hom}_\Lambda(P_1,\Lambda) \to \text{Tr}C \to 0.$$

This notion was first introduced by us at the Stockholm International Congress of Mathematics in 1962 in the context of studying finitely generated modules over regular local rings. It was then used by several people as a means of comparing left and right finitely presented modules over non-commutative rings, as well as in comparing covariant and contravariant finitely presented functors. Thus its introduction to algebra long antidates its use in representation theory. We now describe a basic property of the transpose which underlies its connection with describing the ends of almost split sequences.

Suppose C is an arbitrary Λ-module, A a finitely presented Λ-module. Further, suppose $\mathrm{Tr}A$ is a $\Lambda^{op}-\Sigma$ bimodule. Then the composition of morphisms $\Sigma \to \underline{\mathrm{End}}(\mathrm{Tr}A) \to \underline{\mathrm{End}}(A)^{op}$ makes $\underline{\mathrm{Hom}}(A,C)$ a Σ-module and we have the following basic result which may have a familiar ring to people working with almost split sequences [2].

Proposition 2. For each injective Σ-module I we have an isomorphism

$$\phi : \mathrm{Ext}^1_\Lambda(C,\mathrm{Hom}_\Sigma(\mathrm{Tr}A,I)) \to \mathrm{Hom}_\Sigma(\underline{\mathrm{Hom}}(A,C),I)$$

which is functorial in C and I. Moreover this isomorphism has the following property: If $x = 0 \to \mathrm{Hom}_\Sigma(\mathrm{Tr}A,I) \xrightarrow{f} B \xrightarrow{g} C \to 0$, then $\phi(x) : \underline{\mathrm{Hom}}(A,C) \to I$ has the property that $\mathrm{Im}((\mathrm{Hom}(A,B) \xrightarrow{(A,g)} \mathrm{Hom}(A,C) \to \underline{\mathrm{Hom}}(A,C)) = \mathrm{Ker}(\underline{\mathrm{Hom}}(A,C) \xrightarrow{\phi(x)} I)$.

As an immediate consequence of this proposition we have the following which can often be used to actually construct almost split sequences.

Corollary 3. An exact sequence $0 \to \mathrm{Hom}_\Sigma(\mathrm{Tr}A,I) \to Y \to Z \to 0$ of Λ-modules splits if and only if every morphism $A \to Z$ can be lifted to Y.

It is Proposition 2 which constitutes the starting point for the existence theorems for almost split sequences given here. We begin with the most general existence theorem for almost split sequences for modules over rings [1,2,4].

Theorem 4. Let Λ be an arbitrary ring, C a finitely presented nonprojective Λ-module with End C a local ring and let Γ = End TrC. We consider the unique simple End C^{op}-module a Γ-module by means of the ring surjection which is the composition $\Gamma \to \underline{\text{End}}(\text{TrC}) \overset{\text{Tr}}{\to} \underline{\text{End}}\ C^{op}$ and let I be its Γ-injective envelope. Then we have an almost split sequence

$$0 \to \text{Hom}_{\Gamma}(\text{TrC},I) \to B \to C \to 0.$$

Remark: It should be noted that the uniqueness properties of almost split sequences show that $\text{Hom}_{\Gamma}(\text{TrC},I)$ depends only on C and not on the particular projective presentation of C used in constructing TrC.

We now apply this result to the situation that Λ is an algebra over a complete noetherian local ring R which is a finitely generated R-module. Then both mod Λ and art Λ, the category of artinian modules, are Krull-Schmidt categories and the functor $\text{Hom}_R(\ ,I) : \text{mod}\ \Lambda \to \text{art}\ \Lambda$ gives a duality where I is the R injective envelope of $R/\underline{m}$, and $\underline{m}$ is the maximal ideal of R. Since mod Λ is a Krull-Schmidt category we have that every C in mod Λ has a minimal projective presentation which we use in computing TrC. Finally C in mod Λ or art Λ is indecomposable if and only if End C is local. With these remarks and notations in mind we have the following easy consequence of Theorem 4 [1,2].

Theorem 5. Let C be an indecomposable nonprojective module in mod Λ. Then there is an almost split sequence $0 \to \text{Hom}_R(\text{TrC},I) \to B \to C \to 0$ where $\text{Hom}_R(\text{TrC},I)$ is in art Λ.

Remarks. a) If $\dim R = 0$ (Krull dimension of R), then mod Λ = art(Λ) and the almost split sequence is in mod Λ, and so we obtain one of the original existence theorems of Auslander-Reiten ([4]) which has played an important role in much of the representation theory of artin algebras.

b) If $\dim R > 0$ and depth $R \geq 1$, then $\text{Hom}_R(\text{TrC},I)$ is noetherian, or equivalently of finite length, if and only if pd C = 1. From this it follows that if R is a local ring (complete or not) with depth R ⩾ 1, then the category of modules of finite length has almost split sequences with respect to itself if and only if R is a 1-dimensional regular local ring. We shall have need of this remark when discussing the existence of almost split sequences for coherent sheaves.

Our next existence theorem is a generalization of the one just given. Again we assume that R is a complete local ring with maximal ideal $\underline{m}$ and I the injective envelope of $R/\underline{m}$. We now let $\underline{\underline{P}}$ be an additive R-category in which idempotents split and whose isomorphism classes of objects constitute a set. That is, we assume that $\text{Hom}(P_1,P_2)$ is a finitely generated R-module for all P_1,P_2 in $\underline{\underline{P}}$ and that the composition of morphisms is R-bilinear. For example: R is a field and $\underline{\underline{P}}$ is a quiver with relations over R such that $\text{Hom}(P_1,P_2)$ is of finite dimension for all points P_1 and P_2. We denote by Mod $\underline{\underline{P}}$ the category of all additive contravariant functors from $\underline{\underline{P}}$ to Mod R and we denote by mod $\underline{\underline{P}}$ the full subcategory of Mod $\underline{\underline{P}}$ consisting of all finitely presented functors where $C : \underline{\underline{P}} \to \text{Ab}$ is said to be a finitely presented functor if there is an exact sequence $\text{Hom}_{\underline{\underline{P}}}(\ ,P_1) \to \text{Hom}_{\underline{\underline{P}}}(\ ,P_0) \to C \to 0$ with the P_1 and P_0 in $\underline{\underline{P}}$. As before mod $\underline{\underline{P}}$ is a Krull-Schmidt category so that each C in mod R has a (unique up to isomorphism) minimal projective presentation and C is indecomposable if and only if End C is a local R-algebra. Let $(\ ,P_1) \to (\ ,P_0) \to C \to 0$ be a minimal projective presentation. We define TrC in mod $\underline{\underline{P}}^{op}$ by the exact sequence $(P_0,\ ) \to (P_1,\ ) \to \text{TrC} \to 0$. Then C in mod $\underline{\underline{P}}$ is indecomposable and nonprojective if and only if TrC is indecomposable and nonprojective. Finally for an F in Mod $\underline{\underline{P}}$ we denote by $\text{Hom}_R(F,I)$ the object in Mod $\underline{\underline{P}}^{op}$ which is the functor defined by $\text{Hom}_R(F,I)(P) = \text{Hom}_R(F(P),I)$ for all P in $\underline{\underline{P}}$.

We now have the following generalization of Theorem 5.

Theorem 6. Let C in $\text{mod}\,\underline{\underline{P}}$ be nonprojective and indecomposable. Then we have the almost split sequence

$$0 \to \text{Hom}_R(\text{Tr}C, I) \to B \to C \to 0.$$

We now give examples of such categories $\underline{\underline{P}}$. In these examples we assume that $\dim R = 0$, for instance when R is a field.

a) If $\underline{\underline{P}}$ has the property that $\text{Hom}_R((P\ ,\ ), I)$ and $\text{Hom}_R((\ ,P), I)$ are finitely presented for all P in $\underline{\underline{P}}$ we obtain the notion of a dualizing R-variety [3]. In this case $\text{Hom}_R(\text{Tr}C, I)$ is in $\text{mod}\,\underline{\underline{P}}$ for all C in $\text{mod}\,\underline{\underline{P}}$ and so the almost split sequences $0 \to \text{Hom}_R(\text{Tr}C, I) \to B \to C \to 0$ lie in $\text{mod}\,\underline{\underline{P}}$.

b) If we assume that $(\ ,P)$ and $(P,\ )$ have finite support for each indecomposable P in $\underline{\underline{P}}$, we obtain R-varieties such that $\text{mod}\,\underline{\underline{P}}$ consists of the $\underline{\underline{P}}$-modules of finite length. This includes for instance locally bounded k-categories for a field k [8].

c) Let k be a field and let $\Lambda = \Lambda_0 \amalg \Lambda_1 \amalg \ldots$ be a positively graded 2-sided noetherian k-algebra with the Λ_i finite dimensional k-vector spaces. Denote by $\text{Gr}\,\Lambda$ the category of finitely generated $\mathbf{Z}$-graded Λ-modules with degree zero morphisms. Finally, let $\underline{\underline{P}}$ be the full subcategory of $\text{Gr}\,\Lambda$ consisting of the projective objects in $\text{Gr}\,\Lambda$. Then $\underline{\underline{P}}$ has our desired properties and $\text{mod}\,\underline{\underline{P}}$ is equivalent to $\text{Gr}\,\Lambda$.

We will return to the situation covered by Theorem 6 later on when discussing graded orders.

So far our existence theorems were for almost split sequences which had the almost split properties relative to the whole category of modules or functors. We now give some existence theorems for almost split sequences for lattices over orders which are almost split relative to lattices only and not all modules. There are similar existence theorems for subcategories of $\text{mod}\,\Lambda$ when Λ is an artin algebra which were used in proving existence of preprojective partitions [6], [7]. We begin by recalling the definition of an order and lattices over an order.

Let R be an equidimensional commutative Gorenstein ring of dimension d (Krull dimension). By an R-order we mean an R-algebra Λ which is a finitely generated maximal Cohen-Macaulay R-module having the property that $\mathrm{Hom}_R(\Lambda,R)_{\underline{p}}$ is $\Lambda_{\underline{p}}^{op}$ projective for all nonmaximal prime ideals $\underline{p}$ of R. A Λ-lattice is a finitely generated Λ-module C which is a maximal Cohen-Macaulay R-module such that $C_{\underline{p}}$ is $\Lambda_{\underline{p}}$-projective for all nonmaximal prime ideals $\underline{p}$ of R.

Suppose now that Λ is an R-order where R is a complete local Gorenstein ring of dimension d. Then $\mathrm{mod}\,\Lambda$ is a Krull-Schmidt category. Let C be an indecomposable non-projective in $\underline{\underline{L}}(\Lambda)$, the category of Λ-lattices. Then $\mathrm{Tr}C$ (computed using a minimal projective presentation for C) is indecomposable and nonprojective but not necessarily a lattice. However $\Omega^d\mathrm{Tr}C$ is an indecomposable lattice where $\Omega^d\mathrm{Tr}C$ is defined to be $\mathrm{Ker}\, f_{d-1}$ in a minimal projective resolution $\ldots\to P_d \to P_{d-1} \xrightarrow{f_{d-1}} \ldots \to P_0 \xrightarrow{f_0} C \to 0$ of C. The lattice $\Omega^d\mathrm{Tr}C$ plays the role of the transpose of C in the category of lattices and is denoted by $\mathrm{Tr}_{\underline{\underline{L}}}C$. Also it is not difficult to see that $\mathrm{Hom}_R(\ ,R) : \mathrm{mod}\,\Lambda \to \mathrm{mod}\,\Lambda^{op}$ gives a duality $\mathrm{Hom}_R(\ ,R) : \underline{\underline{L}}(\Lambda) \to \underline{\underline{L}}(\Lambda^{op})$. Keeping these definitions and notations in mind, we have the following [1,2].

Theorem 7. Let C in $\underline{\underline{L}}(\Lambda)$ be indecomposable and nonprojective. Then there is an almost split sequence $0 \to \mathrm{Hom}_R(\mathrm{Tr}_L C,R) \xrightarrow{f} B \xrightarrow{g} C \to 0$ in the category $\underline{\underline{L}}(\Lambda)$, i.e. the terms of the exact sequence are in $\underline{\underline{L}}(\Lambda)$, the sequence does not split and any $X \to C$ in $\underline{\underline{L}}(\Lambda)$ which is not a splittable epimorphism can be lifted to B as well as any $\mathrm{Hom}_R(\mathrm{Tr}_{\underline{\underline{L}}}C,R) \to Y$ in $\underline{\underline{L}}(\Lambda)$ which is not a splittable monomorphism can be extended to B.

As our final general existence theorem we consider graded orders and graded lattices. We begin with some general remarks and notations.

Let k be a field. By a graded k-algebra Λ we mean a k-algebra Λ together with a k-vector space decomposition $\Lambda = \coprod_{i\in\mathbb{Z}} \Lambda_i$ satisfying the following

a) $\dim_k \Lambda_i < \infty$ for all i and $\Lambda_i = 0$ for $i < 0$

b) Λ_0 is semisimple

c) $\Lambda_i = \Lambda_1^i$ and so $\Lambda_i \Lambda_j = \Lambda_{i+j}$

d) Λ is a 2-sided noetherian ring.

By a graded Λ-module M we mean a Λ-module M together with a k-vector space decomposition $M = \coprod_{i \in \mathbf{Z}} M_i$ such that $\Lambda_i M_j \subset M_{i+j}$ for all i and j in **Z**. Moreover with each graded Λ-module M and every n in **Z** we have the shifted graded module M(n) given by $M(n)_d = M_{n+d}$ for all d in Z.

By a graded order Λ over a polynomial ring R = $k[X_1,\ldots,X_n]$ we mean a graded k-algebra Λ together with $X_1,\ldots,X_n$ in Λ_1 algebraically independent elements over k which lie in the center of Λ having the following properties:

a) Λ is a finitely generated free R-module i.e. $\Lambda = \coprod_{i<\infty} R(n_i)$ as an R-module.

b) $\mathrm{Hom}_R(\Lambda,R)_{\underline{p}}$ is $\Lambda_{\underline{p}}$-projective for each graded prime ideal $\underline{p} \neq (X_1,\ldots,X_n)$.

Now the category of finitely generated graded Λ-modules with degree zero morphisms is Krull-Schmidt because $\mathrm{Hom}_{\mathrm{gr}\Lambda}(A,B)$ is finite dimensional over k for all A and B. If we consider $\underline{\underline{L}}(\mathrm{gr}\Lambda)$ the subcategory of graded Λ-modules with degree zero morphisms consisting of all L such that L is R-free and $L_{\underline{p}}$ is $\Lambda_{\underline{p}}$-projective for all graded $\underline{p} \neq \underline{m}$, then we have the following [5].

Theorem 8. If C in $\underline{\underline{L}}(\mathrm{gr}\Lambda)$ is an indecomposable nonprojective lattice, then there is an almost split sequence

$$0 \to \mathrm{Hom}_R(\Omega^n \mathrm{Tr}C, R(-n)) \to B \to C \to 0$$

in the category $\underline{\underline{L}}(\mathrm{gr}\Lambda)$ where Tr and Ω^n are computed using minimal projective presentations and resolutions.

As one consequence of this we have the following [5].

Theorem 9. Let X be a projective curve with only Gorenstein singularities (for instance, X is nonsingular), then the category of coherent sheaves in X has almost split sequences.

Remark: A. Schofield has independently given a sheaf theoretic proof of Theorem 9. Geigle and Lenzing have also obtained some special cases of this result.

References

1. M. Auslander, "Existence theorems for almost split sequences", Proc. Conf. on Ring Theory II, Oklahoma (Marcel Dekker, 1977) pp. 1-44.
2. M. Auslander, "Functors and morphisms determined by objects" Proc. Conf. on Representation Theory (Philadelphia 1976). (Marcel Dekker, 1978) pp. 1-244.
3. M. Auslander and I. Reiten, "Stable equivalence of dualizing R-varieties I", Adv. in Math. 12 (1974) 306-366.
4. M. Auslander and I. Reiten, "Representation theory of artin algebras III. Almost split sequences", Comm. Algebra 3 (1975) 239-294.
5. M. Auslander and I. Reiten, "Almost split sequences for graded modules",
6. M. Auslander and S.O. Smalø, "Preprojective modules over artin algebras", J. Algebra 66 (1980) 61-122.
7. M. Auslander and S.O. Smalø, "Almost split sequences in subcategories", J. Algebra 69 (1981) 426-454.
8. K. Bongartz and P. Gabriel, "Coverings in representation theory", Invent. Math. 65 (1982) 331-378.

REPRESENTATIONS OF POSETS AND TAME MATRIX PROBLEMS

A.V. Roiter
Institute of Mathematics, AN USSR, Ul. Repina 3, 252601 Kiev/USSR

§ 1. The theory of matrix problems can be considered in a natural way as a part or prolongation of homological algebra (see [1]). Let K be an arbitrary category. Beside the notations $\mathrm{Obj}\,K$, $\mathrm{Mor}\,K$, $\mathrm{Ind}\,K$ and $K(A,B)$ generally used for the collections of objects, of morphisms, of isoclasses of indecomposables and of morphisms from A to $B (A,B \in \mathrm{Obj}\,K)$, we shall also use the notation $\mathrm{El}\,K$ for the collection of all elements of all objects of K; this makes sense whenever K is a concrete category, i.e. a subcategory of the category of sets.

Let Φ be a functor with values in a concrete category V. To be precise, suppose that Φ depends on two variables belonging to categories K_1 and K_2 and that it is contravariant in the first and covariant in the second. Then $\mathrm{El}\,\Phi$ will denote the collection of all elements of all sets $\Phi(A,B)$, where $A \in \mathrm{Obj}\,K_1$ and $B \in \mathrm{Obj}\,K_2$. With this rather strange terminology, we can say that one of the fundamental ideas of homological algebra consists in the remark that some categories G (of groups, algebras, modules...) can be investigated by means of a naturally constructed map h from $\mathrm{El}\,\Phi$ to $\mathrm{Obj}\,K$, where Φ happens to be one or another appearance of the functor Ext. The map h is surjective if the objects of G are considered up to isomorphisms, but it is far from being injective. Therefore, in order to get a practical classification of the objects of G, it is necessary to make clear what elements of the functor Ext really correspond to isomorphic objects of G. For this, we must consider $\mathrm{Ext}(A,B)$ as a group on which $K_1(A,A)$ and $K_2(B,B)$ act. And in this context it is natural to attach a category $\hat{\Phi}$ to an arbitrary functor Φ in such a way that h embeds into a commutative diagram

$$\begin{array}{ccc} \mathrm{El}\,\Phi & \longrightarrow & \mathrm{Obj}\,\hat{\Phi} \\ {\scriptstyle h}\searrow & & \swarrow{\scriptstyle \bar{h}} \\ & \mathrm{Obj}\,G & \end{array}$$

The so obtained map $\bar{h}$ is not always surjective, but it is so in many important cases. And in all cases it is much closer to being injective than h.

In order to construct $\hat{\Phi}$, we set $\mathrm{Obj}\,\hat{\Phi} = \mathrm{El}\,\Phi$ and $\hat{\Phi}(x,y) = \{(\varphi,\psi) \mid \varphi : A_1 \to A_2, \psi : B_1 \to B_2, x\Phi(e,\psi) = y\Phi(\varphi,e)\}$ for all $x \in \Phi(A_1,B_1)$ and $y \in \Phi(A_2,B_2)$ (we write the operators on the right).

Many (but not all!) categories arising from matrix problems are in fact categories of the form $\hat{\Phi}$, and precisely this explains why matrix problems can be applied to various mathematical questions and in particular to the theory of representations of finite-dimensional algebras.

We shall assume that V is the category of all finite-dimensional vector spaces over an algebraically closed field k and that K_1, K_2 are completely additive categories over k. This means that each idempotent of K_i has a kernel (K_i is a Krull-Schmidt category in the sense of [2]), that the spaces $K_i(A,B)$ are finite-dimensional vector spaces and that the composition of K_i is k-bilinear. In fact, in many cases, one of the categories K_i will be given some special form.

In particular, the study of the representations of an arbitrary finite-dimensional algebra can be reduced to the investigation of some category $\hat{\Phi}(K_1,K_2)$, where K_1 or K_2 is <u>simple</u>, i.e. has (up to isomorphisms) only one indecomposable object whose ring of endomorphisms is identified with k. This is of course due to the fact, that any module can be represented as an extension of some submodule (resp. quotient) by a quotient (resp. submodule) which is a direct sum of isomorphic simple modules. This example shows how essential from a homological point of view is the case where one of the categories, say K_2, is simple. In this case, the functor Φ is completely determined by the functor Φ_1 of one variable which is defined on K_1 by $\Phi_1(A) = \Phi(A,U)$, where U is an indecomposable object of K_2; if $x \in K_1(A_1,A_2)$, we set $\Phi_1(x) = \Phi(x,e)$, where e is the identity of U.

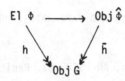

Furthermore, in this case we can essentially suppose without loss of generality that Φ_1 is an embedding of K_1 into V and that $\hat{\Phi}_1$, which we then also denote by $\hat{K}_1$, is defined by a subcategory K_1 of V. Notice that in this case $\hat{K}_1$ does "almost not differ" from the subspace category produced by K_1 (see [3], [4]; in fact, it differs by a finite number of "trivial" indecomposables). For various further limitations imposed on K_1, the categories $\hat{K}_1$ will be the fundamental objects of our investigation. Thereby we agree to assume that K_1 has only finitely many indecomposables (although, generally speaking, this is not essential).

Notice that in K_1 each object A and each space of morphisms has a dimension (over the field k). The notion of a basis and of a multiplicative basis of K_1 is defined in a natural way (compare with [12]).

To be precise, let $A_1,\ldots A_n$ be representatives of the indecomposable objects of K_1; we call <u>basis</u> of K_1 a family of bases of all spaces A_i together with a family of bases of all morphism spaces $K_1(A_i,A_j)$. The basis is called <u>multiplicative</u> if, for each basis morphism $x \in K_1(A_i,A_j)$ and each basis vector $a \in A_j$, the product $a\Phi_1(x)$ is either a basis vector of A_i or zero. The basis is called <u>weak</u> if each basis morphism different from an identity has rank 1 (as a linear operator).§ The category K_1 will be called 1) a <u>poset category</u>, 2) <u>schurian</u>, 3) <u>weak</u>, 4) <u>a category with multiplicative basis</u>, if 1) $\dim I = 1$ for all $I \in \operatorname{Ind} K_1$, 2) $\dim K_1(I,I) = 1$ for all $I \in \operatorname{Ind} K_1$, 3) K_1 has a weak multiplicative basis, 4) K_1 has a multiplicative basis.

In the case 1), the category K_1 is completely determined by some poset S, and $\hat{K}_1$ is the category $R(S)$ of representations of S in the sense of [4]. It is easy to see that 2), 3) follow from 1), and of course 4) follows from 3). Besides, we can as usual define the notions of <u>finite</u> and <u>tame type</u> for K_1 (or more precisely for $\hat{K}_1$). It is not difficult to prove that 1) and 2) coincide in the finite type case. It happens that multiplicative bases then always exist, but we will not discuss this question here.*

---

* In [3] a proof of a weaker result is essentially given: In the finite type case there is always a "semimultiplicative basis (compare with [12]).

§ and if non-zero products of different basis vectors by the same basis morphism are different.

So the different possible cases can be displayed as follows:

	1	2+3	2+4	2	3	4	A⌐B
F							
T							

The representations of posets of finite type are examined in [4],[5]. We shall recall the results in § 2. In case F_3, the result is obtained in [3],[6]. In case F_1, the fundamental criterion is given in [7]. in § 3 we shall formulate a generalization of this result to T_2.

§ 2. Posets of finite type

Considering that $\Phi(\oplus_{i=1}^{m} I_i^{k_i}, \oplus_{j=1}^{n} J_j^{l_j}) = \oplus_{i,j} \Phi(I_i, J_j)^{k_i l_j}$, we can interprete each object of $\hat{\Phi}$, i.e. each element of Φ, as a matrix whose coefficients belong to the spaces $\Phi(I,J)$, where $I \in \operatorname{Ind} K_1$ and $J \in \operatorname{Ind} K_2$. In the case where the spaces $\Phi(I,J)$ have dimension 1, we can assume that the coefficients of the matrix belong to the field k.

In particular, if K_1 is a poset category (and K_2 is simple), each object of $\hat{K}_1$ is identified with a matrix divided into n vertical bands indexed by the elements of the poset S of cardinality n. Two such matrices are isomorphic as objects of $\hat{K}_1$ if one can be obtained from the other by a sequence of elementary transformations of the following types:

1) arbitrary transformations of the rows;
2) arbitrary transformations of the columns within each band;
3) addition of a column of the band s_i to a column of the band s_j if $s_i < s_j$ (see [4]).

Beside representations of a poset, it is natural to consider also matrices divided into horizontal bands as well as into vertical ones; the transformations of the columns are now governed by some poset S, those of the rows by a poset T. As it seems natural, the matrix occuring in this problem will be called a <u>representation of the pair of posets</u> S <u>and</u> T [5]; we will denote by $R(T \wr S)$ the category of these representations.

In the categorial language, the representations of a pair of posets are the objects of a category $\hat{\Phi}$, where K_1 and K_2 are poset categories and $\Phi(I_i, I_j)$ has dimension one for all $I_i \in \mathrm{Ind}\, K_1$ and $J_j \in \mathrm{Ind}\, K_2$.

A fundamental method in the investigation of matrix problems consists in constructing various algorithms which reduce one problem to another. For instance, consider a representation of a poset consisting of two noncomparable elements, i.e. a matrix $(A|B)$. Reduce A to the form $\begin{pmatrix} E & 0 \\ 0 & 0 \end{pmatrix}$ and perform transformations on B which do not change the form obtained for A . In this way, we are led to the problem of a matrix B divided into two horizontal bands such that the rows of the lower band can be added to those of the upper one. So, after transposition, we are led to the representation problem of a poset consisting of two elements which now are comparable. We express this by the "formula" $R(1,1) \Rightarrow R(2)$.

More generally, we denote by (n) a chain of n points, by $(k_1,\dots,k_t)$ the <u>primitive</u> poset (in the sense of [4]) formed by t disjoint chains of cardinality k_i , where elements of distinct chains are incomparable. We shall write $X \Rightarrow Y$ if the category X can be "reduced" to the category Y . More precisely, this means that we can construct a bijection between a cofinite subset of $\mathrm{Ind}\, X$ and all of $\mathrm{Ind}\, Y$. In general the representations of $\mathrm{Ind}\, X$ excluded by the construction will be trivial in some sense.

Almost as trivially as $R(1,1) \Rightarrow R(2)$, we can prove that

(1) $\qquad R((n),T) \Rightarrow R((n+1)\wr T)$

where $((n),T)$ is the disjoint sum of two non-comparable posets (n) and T . Notice that (1) already permits to reduce the question of the finiteness of the type of a poset pair to the corresponding question for one poset [5]. Indeed, it is easy to see that $R((1,1)\wr(1,1)$ is of infinite type, since it is equivalent to the category of representations of the quiver  . As a consequence, if $R(S\ T)$ is of finite type, S or T must be a chain, so that we can use (1).

We call <u>width</u> $w(s)$ of an element $s \in S$ the maximal cardinality of an <u>antichain</u> (= subposet of the form $(1,1,\dots,1)$) of S containing s . The <u>width</u> $w(S)$ of S is the supremum of the widths of its elements.

We call <u>semichain</u> a poset G of width 2 which admits no subset of the form (1,2) (for the induced order). In § 3 we shall need the following generalization of (1)

(2) $R(G,T) \Rightarrow R(\bar{G}\ T)$,

where $\bar{G}$ is the semichain obtained from G by addition of a finite number of points of width 1 (more precisely, $\bar{G}$ is the complement of the maximal element in the modular lattice generated by G).

For each maximal element $a \in S$ we construct a new poset S'_a by eliminating a from S and adding a new element $b \cup c$ whenever $b,c \in S$ are incomparable with a and between themselves. It is proved in [4] in the matrix language and in [9] in the vector space language that

(3) $R(S) \Rightarrow R(S'_a)$

if $w(S) \leq 3$. Since posets of width 4 are known to be of infinite type (due to the matrices $\left(\begin{smallmatrix} E \\ 0 \end{smallmatrix} \middle| \begin{smallmatrix} 0 \\ E \end{smallmatrix} \middle| \begin{smallmatrix} E \\ E \end{smallmatrix} \middle| \begin{smallmatrix} E \\ J \end{smallmatrix}\right)$, where J is a Jordan-block), (3) reduces the study of posets of finite type to combinatorics.

Since "differentiation" decreases the number of indecomposable representations, the posets of finite type "come down to nothing" after several differentiations, i.e. they are transformed into the empty set. This is completely obvious for posets of width 1 or 2 , because $S'_a = S \setminus a$ if $w(S) \leq 2$.

As for posets of width 3, they may be of finite as well as of infinite type.

<u>Theorem of Kleiner.</u> S <u>is of finite type if and only if it contains no subset of the following form</u> (<u>for the induced order</u>):

1) (1,1,1,1)
2) (2,2,2)
3) (1,3,3)
4) (1,2,5)
5) N_4 = $\amalg$ (4)

<u>Necessity.</u> We already know that (1,1,1,1) is of infinite type.

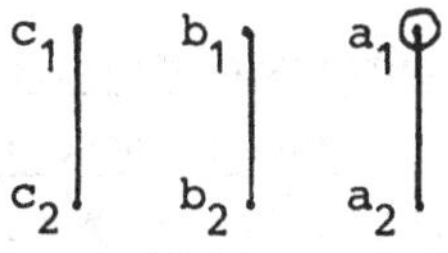

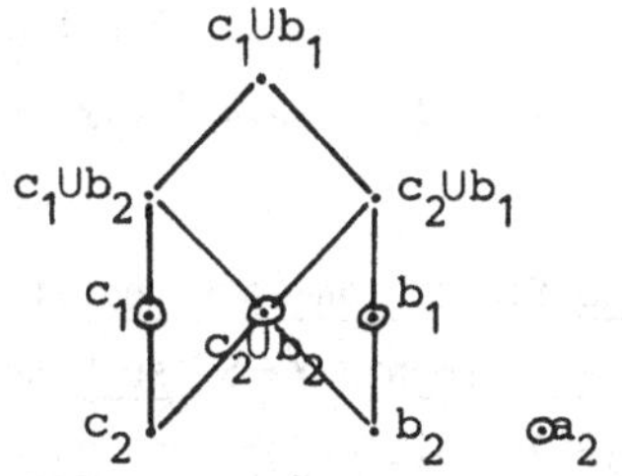

$S = (2,2,2)$ $\qquad$ $S'_{a_1} \supset (1,1,1,1) = (c_1, c_2 \cup b_2, b_1, a_2)$

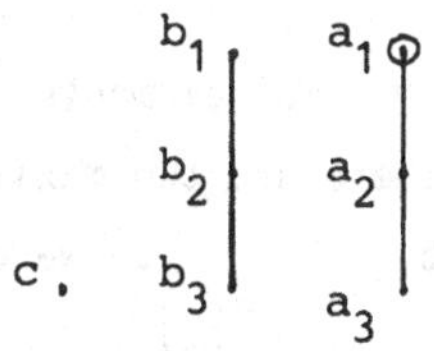

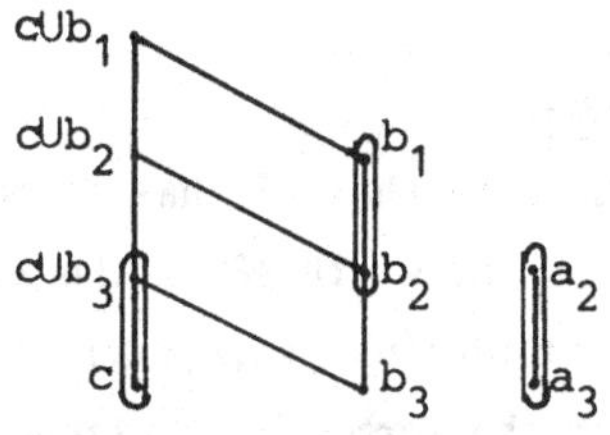

$S = (1,3,3)$ $\qquad$ $S'_{a_1} \supset (2,2,2)$

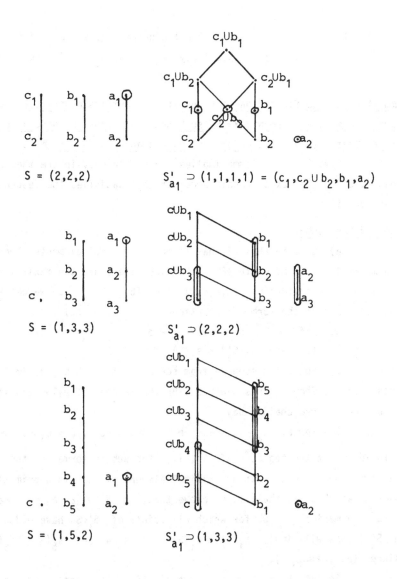

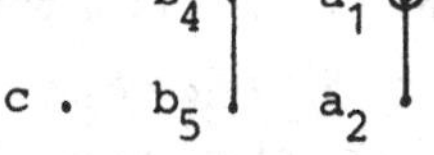

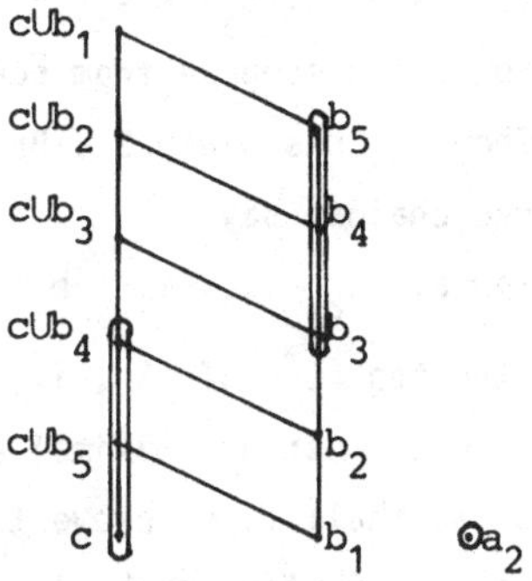

$S = (1,5,2)$ $\qquad$ $S'_{a_1} \supset (1,3,3)$

On the other hand, $((1,5,2)'_{b_1})'_{a_1 \cup c} = N_4$.

In order to prove the <u>sufficiency</u>, we have to show that, if S contains no "critical" subset 1),2),3),4) or 5), then S'_a contains none and is even better than S in some sense. For that, we shall compare the primitive subsets of S and S'_a . More precisely, to each S we attach the following invariant $s(S) \in \mathbb{N}^6$: In case $i = 1,3$ and 4 , $s_i(P)$ is the number of subsets of S of the form $(1,2,4)$, $(1,2,3)$ and $(1,2,2)$

respectively; $s_2(S)$ and $s_5(S)$ are the numbers of subsets of the form $N_3 = \mathbb{N}\,\mathbb{I}$ and (1,1,1) respectively; finally, $s_6(S)$ is the cardinality of S .

Lemma 1. If the finite non-empty poset S contains no critical subset, there is a maximal element $x \in S$ such that S'_x contains no critical subset and that $s(S'_x) < s(S)$ for the lexicographic order of $\mathbb{N}^6$.

Clearly, this lemma implies the sufficiency in the theorem of Kleiner, since the lexicographic order of $\mathbb{N}^6$ satisfies the descending chain condition.

Proof of the lemma.

a) Suppose that S has at most 2 maximal elements. Then there is a maximal $a \in S$ of width ≤ 2 : Otherwise, there are two maximal elements $a,b \in S$ and two antichains $\{a,a_1,a_2\}$ and $\{b,b_1,b_2\}$. So we get a contradiction in the form of an antichain $\{a_1,a_2,b_1,b_2\}$.

Now, if $a \in S$ is maximal of width ≤ 2 , S'_a equals $S \setminus \{a\}$. So $s_6(S'_a) < s_6(S)$ and $s_i(S'_a) = s_i(S)$ for $i \leq 5$.

b) So we can suppose from now on that S has 3 maximal elements a,b,c . Then S has width 3 . By the theorem of Dilworth, it is the union of three chains, say

$$S = \{a = a_1 > a_2 > \ldots > a_l\ ,\ b = b_1 > \ldots > b_m\ ,\ c = c_1 > \ldots > c_n\}$$

It follows that the top S^+ of S , i.e. the subset formed by the elements smaller or equal to at most one maximal element of S , is a primitive poset of width 3. We shall first prove that, if $S^+ = \{a > a_2 > \ldots > a_g, b, c\}$, there is a maximal $x \in S$ for which all points of $S'_x \setminus S$ have width ≤ 2 in S'_x . This will imply $s_i(S'_x) \leq s_i(S)$ for $i \leq 4$ and $s_5(S'_x) < s_5(S)$ (since $\{a,b,c\} \not\subset S'_x$).

So let $S^+ = \{a_1 > \ldots > a_g, b, c\}$. If $a > b_i$ for $i \geq 2$, $b \cup c$ is the only "new" element in S'_a and it has width ≤ 2 . So we are left with the case where $b_2 \not< a$ and $c_2 \not< a$, hence $b_2 < c$ and $c_2 < b$. The new elements of S'_c are then $a_1 \cup b, \ldots, a_g \cup b$. They obviously have width ≤ 2 in S'_c .

c) By a) and b) we can now suppose that $S^+ = \{a_1 > \ldots > a_g, b, c_1 > \ldots > c_h\}$ with $b < a_1$, $\inf(g,h) = 2$ and $\sup(g,h) \leq 4$. It follows that

$$S_a' = \{a_2 > .. > a_1 , b \cup c_1 > .. > b \cup c_h > b_1 > .. > b_m , c_1 > .. > c_n\} .$$

Let further P be a maximal primitive subset of S_a' which has width 3 and contains some $b \cup c_j$. Then P intersects the three exhibited chains of S_a' in three intervals A, B and C . We must have $k < j$ for each $c_k \in C$, because c_k is incomparable with $b \cup c_j$. Moreover, $b = b_1 \in B$ because P is maximal (otherwise, b would be comparable with some $x \in A$ and $b \cup c_j > b > x$). For the same reason, $c_{j-1} \in C$ if $b \cup c_j$ is maximal in B . So P has the form given below, and S contains the primitive subset $P^{-1} = A \cup \{b_1, \ldots, b_q\} \cup C \cup \{c_{h-r+1}, \ldots, c_h\}$.

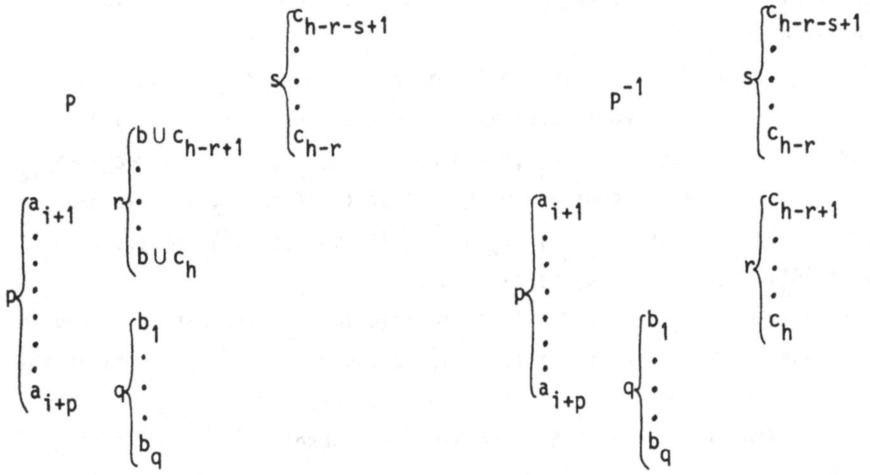

Now, we claim that S_a' contains no subset T of the form (2,2,2) , or equivalently that $p = 1$ or $s = 1$: Otherwise,

$$\{a_1 > a_{i+1} > a_{i+2} , b , c_{h-r-1} > c_{h-r} > c_h\}$$

would be a subset of S of the form (1,3,3) for some P containing T .

Similarly, S_a' contains no subset of the form (1,3,3). Otherwise, we would have $q+r \geq 3$ and $p \geq 3$ or $s \geq 3$ for some P . In case $p \geq 3$, P^{-1} would contain the primitive subset
$\{a_{i+1} > a_{i+2} , b_1 > b_2 , c_{h-r} > c_h\}$ if $q \geq 2$ and the subset
$\{a_{i+1} > a_{i+2} > a_{i+3} , b_1 , c_{h-r} > c_{h-1} > c_h\}$ if $r \geq 2$. In case $s \geq 3$, S would contain $\{a_1 > a_{i+1} , b_1 , c_{h-r-2} > c_{h-r-1} > c_{h-r} > c_{h-1} > c_h\}$ if $r \geq 2$

and $\{a_{i+1} < a_1 > b_2 < b_1, c_{h-3} > c_{h-2} > c_{h-1} > c_h\} \stackrel{\sim}{\to} N_4$ if $r = 1$.

d) We now examine the case where P is isomorphic to $(1,2,k)$ and k is maximal. In case $k = 1$, we have $s_t(S_a') \leq s_t(S)$ for $t \leq 4$ and $P = \{a_2, b_1 \cup c_h > b_1, c_{h-1}\}$, hence $s_4(S_a') < s_4(S)$ because $P^{-1} \cup \{a\} \not\subset S_a'$. It follows that $s(S_a') < s(S)$ and that we are reduced to the case $k \geq 2$.

Since P^{-1} is a primitive subset of S, we then have $q = 1$ or $p = 1$. In case $q = 1$, $P^{-1} \cup \{a\}$ is a primitive subset of S of the form $(p+1, 1, r+s) = (k+1, 1, 2)$ or $(2, 1, k+1)$. This implies $k \leq 3$ and $s_1(S_a') \leq s_1(S)$. In particular, S_a' contains no subset of the form $(1,2,5)$ or N_4. If $k = 3$, we have $s_1(S_a') < s_1(S)$ because of $P^{-1} \cup \{a\}$, hence $s(S_a') < s(S)$. If $k = 2$, we have $s_2(S_a') \leq s_2(S)$ and $s_3(S_a') < s_3(S)$, hence again $s(S_a') < s(S)$.

e) We are left with the case $p = 1, q = s = 2$ (in case $q \geq 3$ or $s \geq 3$, P^{-1} or P would contain a subset of the form $(1,3,3)$). Then S contains the subset $N = \{a_{i+1} < a_1 > b_2 < b_1, c_{h-r-1} > c_{h-r} > \ldots > c_h\} \stackrel{\sim}{\to} N_{r+2}$, and $r = 1$ because S contains no subset of the form N_4. So we have $s_1(S_a') \leq s_1(S)$. In case $s_2(S_a') < s_2(S)$, this implies $s(S_a') < s(S)$. In case $s_2(S_a') \geq s_2(S)$, S_a' contains a subset $N' = \{x_2 < x_1 > y_2 < y_1, z_1 > z_2 > z_3\}$ isomorphic to N_3 and not contained in S (because $N \not\subset S_a'$). We claim that $y_1 \in S$: otherwise, S_a' contains the subset $\{x_1 > x_2, y_1 > b, z_1 > z_2\}$ isomorphic to $(2,2,2)$.

Now, if $x_1 \notin S$, S_a' contains the subset $P' = \{x_1 > x_2, y_1, z_1 > z_2 > z_3\}$ for which the number q' corresponding to the invariant q of P equals 1. By d) it follows that $s_1(S_a') < s_1(S)$, hence $s(S_a') < s(S)$.

Finally, if $z_1 \notin S$, we set $P' = N' \setminus \{y_2\}$ or $P' = N' \setminus \{x_1\}$ so as to obtain a primitive subset for which the number p' corresponding to the invariant p of P equals 2. Then again we have $s_1(S_a') < s_1(S)$ by d) and $s(S_a') < s(S)$.

§ 3. Tame schurian categories

The fundamental aim of this paragraph is to formulate and explain the criterion recently obtained by L.A. Nazarova which divides the schurian categories into tame and wild ones.

In this paragraph we consider the category $\hat{K}_1$ where K_1 is a schurian subcategory of the category V ($\dim K_1(I,I) = 1$, $I \in \operatorname{Ind} K_1$). Whenever we speak of objects of K_1, we always mean indecomposable ones.

An object I is called small or large according as $\dim I = 1$ or $\dim I \geq 2$. If $\operatorname{Ind} K_1$ contains only one object I, $\hat{K}_1$ is the category of representations of a quiver with two points and k arrows (from the first point to the second; $k = \dim I$). Accordingly, as we already noticed in § 1, all objects are small if $\hat{K}_1$ is of finite type; this means that K_1 is a poset category. It is clear that $\dim I \leq 2$ if $\hat{K}_1$ is tame. Therefore, we shall assume that all large points have dimension 2.

We also divide the morphisms (between indecomposables of K_1) into small and large ones, calling a morphism small if its rank as a linear operator equals 1. We call a subcategory $\bar{K}_1 \subset K_1$ semifull if, for all $I_i, I_j \in \operatorname{Ind} \bar{K}_1$, $\bar{K}_1(I_i,I_j)$ contains all small morphisms of $K_1(I_i,I_j)$.

If $\bar{K}_1$ is a weak schurian category (see § 1), we call points of $\bar{K}_1$ the basis vectors of the objects in a chosen weak multiplicative basis. A point is called small (notation $\cdot$) or large (notation $\circ$) if it lies in a small or a large object respectively. Large points belonging to the same object are called connected; sometines we connect them by a dotted line. All large and small points form a poset $S(\bar{K}_1)$.

Now we formulate the wildness criterion: The category K_1 is wild if and only if it contains a wild weak semifull subcategory $\bar{K}_1$. A weak category $\bar{K}_1$ is wild if and only if $S(\bar{K}_1)$ contains one of the following subsets (endowed with the order induced by $S(\bar{K}_1)$.

1) (2,1,1,1) ; 1') 1,1,1,1,1) ;
2) (3,2,2)
3) (4,3,1)
4) (6,2,1)
5) N_5 = 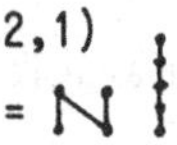
6) $\circ\ \circ\ \cdot$; 6') $\circ\ \circ\ \circ$
7) $\circ\ \cdot$ ⁝ ; 7') $\circ\ \cdot\ \cdot\ \cdot$

If this is not the case, $\bar{K}_1$ is tame.

The sets 1) - 5) are those occuring in the wildness criterion for poset categories [7]. In the schurian case, the proof follows the same plan as in the poset case; in some sense, it is even simpler because schurian categories are more natural in the tame case than posets.

We remark that we do not draw dotted lines in 6,6',7,7' intentionally, i.e. we do not make precise what large points belong to the same object. For instance, 6) ○ ○ • may denote the category ○--○ • or part of the category

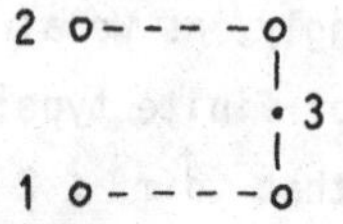

Accordingly, when we consider 6 and 7 in the proof of the necessity, we must examine several weak subcategories containing these pictures and even non-weak categories which contain these weak categories as half-full subcategories. In particular, the category K_1 is of wild type if it consists of two large objects $I, \bar{I}$ with a two-dimensional morphism-space $K_1(I,\bar{I})$ $(K_1(\bar{I},I) = 0)$ and has a (non-weak) multiplicative basis $e_1, e_2 \in I$, $\bar{e}_1, \bar{e}_2 \in \bar{I}$, $\varphi, \varphi_{12} \in K_1(I,\bar{I})$ such that $e_1\varphi = \bar{e}_1$, $e_2\varphi = \bar{e}_2$, $e_1\varphi_{12} = \bar{e}_2$, $e_2\varphi_{12} = 0$. Indeed, K_1 contains the half-full subcategory $\bar{K}_1$ obtained from K_1 by deletion of φ, having the form given below and therefore containing 6'.

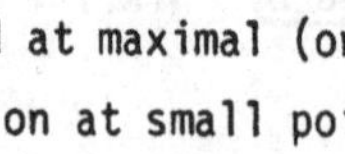

For the proof of the sufficiency, the first thing is to generalize the algorithm of differentiation examined in the preceding paragraph to schurian categories. We notice that, whereas "previously" differentiation was permitted at maximal (or minimal) points of width ≤ 3, now we allow differentiation at small points which have width ≤ 4 and do not occur in • • ○ .

As previously, the point a vanishes in S'_a and new small points $b \cup c$ appear whenever a,b,c are not comparable. But now, if a occurs in a quadruplet of four small points $abcd$, we also have to add a small point $b \cup c \cup d$ and two large points in the following way

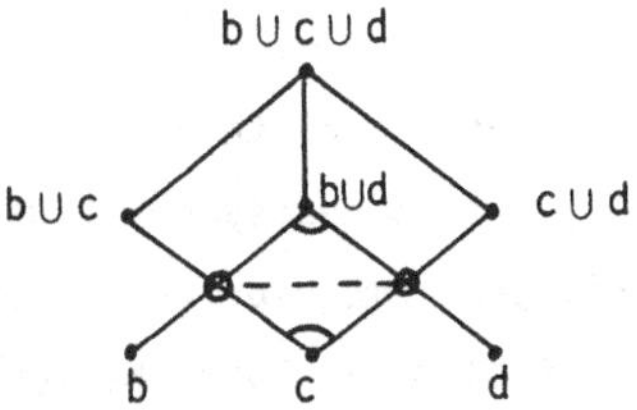

Accordingly, the large new points arise at the intersections of some lines of the "old" picture

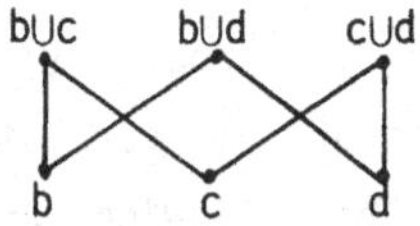

The so constructed category has no multiplicative basis. By two lines connected by an arc 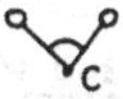 we denote a morphism φ such that $c\varphi = e_1+e_2$. By  we denote a morphism ψ such that $e_1\psi = e_2\psi$ (e_1, e_2 are the basis vectors of the arising large object).

Moreover, if a was comparable with one only out of two "connected" large points 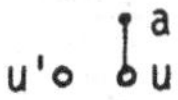, then a new small point u^+ arises which satisfies the same inequalities as u with respect to the other elements. In other words, we get .

We notice that the differentiation does not require the existence of a multiplicative basis in K_1. But it is convenient to suppose that K_1 has a weak multiplicative basis "with respect to the point a" at which the differentiation takes place. This means that the restrictions imposed on a weak multiplicative basis are satisfied for all basis morphisms with range a, a condition which can always be met.

In this way, we can now differentiate a quadruplet of incomparable (small) points, verify trivially that the problem is tame (reducing it to the pencil) and list all the indecomposable representations [9],[10].

Let us now produce the plan of the proof of the sufficiency. We restrict to a weak category with poset of finite growth[11]. Take such a category $\bar{K}_1$ not containing 1 - 7'. In analogy with [7] it can be proved without great difficulties that if our category contains no subset 1-7' (in the sense explained above) before differentiation, then it contains none either after differentiation. As in [7], it is necessary to use the reduction to the so-called Gelfand matrix problems, for which tameness is proved in [11]. As an example of such problems we have $R(G_1 \wr G_2)$, where G_1, G_2 are semichains. Accordingly, if a poset is the union of two semichains, then formula (2) reduces it to the Gelfand problem, which implies that it is of tame type. This procedure is extended to schurian categories. A subset G of width 2 of the set $S(\bar{K}_1)$ (with large and small points) is called semichain if it does not contain •] and • o . The union of two semichains can then be reduced to the Gelfand problem (in the sense of [11]; it can no longer be interpreted as a category $\hat{\Phi}$ for some functor Φ). The third used method is "dislocation". The following non-trivial lemma holds.

Lemma 2. Let $S(\bar{K}_1)$ be decomposed as the union of disjoint subsets U, V, W , where $u < w$ for arbitrary $u \in U, w \in W$, and where no $v \in V$ is smaller than some $u \in U$ or larger than some $w \in W$. Let all points of V be small and all pairs of connected points be contained entirely either in U or in W . Finally, suppose that $S(\bar{K}_1)$ contains no subset

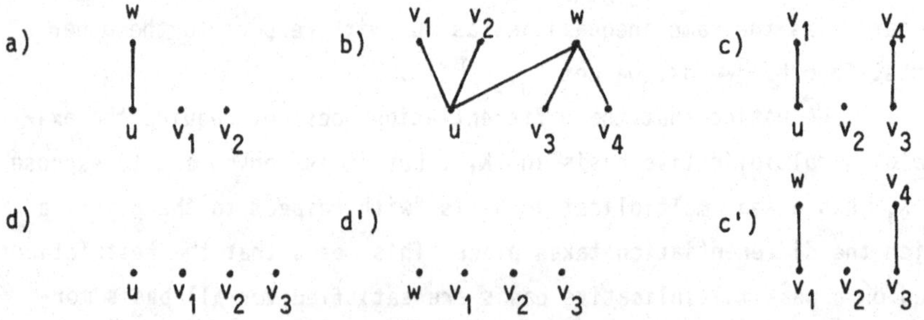

where the points u, w may be large. Then, an arbitrary indecomposable representation of $\bar{K}_1$ can be realized in $U \cup V$ or in $V \cup W$.

Accordingly, in order to prove the sufficiency of the criterion, we have to show that an arbitrary non-differentiable category which contains no subset 1 - 7' is either dislocated or the union of two semichains. Let us produce here the proof of this fact for a weak category $\bar{K}$, for which we suppose moreover (to shorten the argument) that there are no distinct objects $A,B \in \bar{K}$ such that $K(A,B) \neq 0$ and $\bar{K}(B,A) \neq 0$.

Lemma 3. $N(\circ)$ and $N(\cdot\,\cdot)$ are semichains.

If $T \subset S$, $N(T)$ denotes $\{x \in S \mid \forall y \in T,$ x and y are not comparable$\}$. Lemma 3 follows from the absence of 1, 6, 7, 7' in $S(\bar{K}_1)$. Lemma 4 is trivial;

Lemma 4. If the intersection of two semichains equals $(\circ)$ or $(\cdot\,\cdot)$ and if this intersection is maximal in one semichain and minimal in the other, then the union of these semichains is again a semichain.

Under the given restrictions, all large objects $A_1,\dots,A_n$ can be numbered in such a way that $\bar{K}(A_i,A_j) \neq 0$ iff $i<j$. By A_i we also denote the pair of points of $S(\bar{K})$ belonging to A_i , by $A_i^-(A_i^+)$ the set of points which are smaller (larger) than one element of A_i at least. Each small point either falls to $A_i^+ \cap A_{i+1}^-$ for some i , or to A_1^- or to A_n^+ .

Lemma 5. If the sets $U_i = A_i^- \cup A_i$, $V_i = A_i^+ \cap A_{i+1}^-$ and $W = A_{i+1} \cup A_{i+1}^+$ do not satisfy the conditions of lemma 2, then $V_i \cup A_i \cup A_{i+1}$ is the union of two semichains.

Under the given assumptions, we can suppose that $u \in A_i$, $w \in A_{i+1}$. Therefore, the conditions c,c',d,d' follow from the absence of 6 and 7 in $S(\bar{K})$.

Accordingly, the conditions of lemma 2 can fail to be satisfied for three reasons:

1) One of the points $a_i, a_i' \in A_i$ is not comparable with one of a_{i+1}, $a_{i+1}' \in A_{i+1}$;
2) condition a is not satisfied;
3) condition b is not satisfied.

In the first case $V_i \cup A_i \cup A_{i+1} = N(a_i) \cup N(a_{i+1})$ $(a_i < x < a_{i+1} \rightarrow a_i < a_{i+1})$.

In the second case, there exist points v_1, v_2 which are not comparable with a_i, a_{i+1}

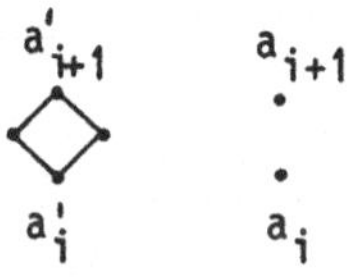

$G_1 = N(v_1) \cap N(v_2)$ is a semichain by lemma 3.

Let $G_2^-(G_2^+)$ be the set of points smaller (larger) or equal to at least one of the points v_1, v_2. Then $G_2^- \subset N(a_i)$, $G_2^+ \subset N(a_{i+1})$. It follows that G_2^- and G_2^+ are semichains and that $G_2 = G_2^- \cup G_2^+$ is a semichain by lemma 4. $V_i \cup A_i \cup A_{i+1} = G_1 \cup G_2$.

Suppose finally that the following condition is not satisfied:

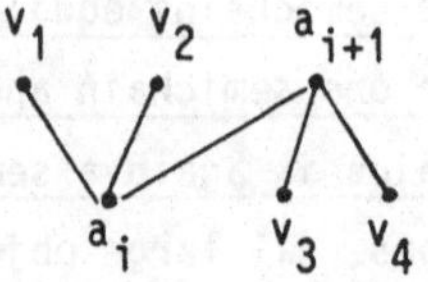

We set $G_1^- = (v_1,v_2)^- \setminus ((v_1,v_2)^- \cap (v_3,v_4)^-)$, $G_1^+ = (v_1,v_2)^+$, $G_2^- = (v_3,v_4)^-$, $G_2^+ = (v_3,v_4)^+ \setminus ((v_3,v_4)^+ \cap (v_1,v_2)^+)$. Then $G_1^- \subset N(v_3,v_4)$, $G_1^+ \subset N(a_{i+1})$, $G_2^- \subset N(a_i)$, $G_2^+ \subset N(v_1,v_2)$. Accordingly, $G_1 = G_1^- \cup G_1^+$ and $G_2 = G_2^- \cup G_2^+$ are semichains (by lemma 3 and 4).

Lemma 6. If $\bar{K}$ is differentiable at no maximal (minimal) point, $A_n^+ \cup A_n$ $(A_1^- \cup A_1)$ is a semichain.

In fact, if $\bar{K}$ is differentiable at no maximal point, there is no maximal small point comparable with both points of A_n. This means that $A_n^+ \cup A_n \subset N(a_n) \cup N(a_n')$.

In order to complete the proof of the fact that $S(\bar{K})$ is the union of two semichains, it remains to match the semichains obtained in lemma 5 and 6 according to lemma 4.

Finally, we report that V.M. Bondarenko, L.A. Nazarova and the author have also obtained a finite growth criterion for schurian categories. This generalizes the analogous criterion for posets [11]. We formulate this criterion here for a weak category $\bar{K}$: Such a $\bar{K}$ is of infinite growth if $S(\bar{K})$ contains • • $\bowtie$ or if it containes a subset T satisfying the following conditions: Each large point of T is connected to another point of T, and each large point has width 3 or is not comparable with a large point of T not connected with it.

The author thanks P. Gabriel for the translation of this article.

Literature

[1] A.V. Roiter, Matrix problems and representations of bisystems, Zap. Naučn. Sem. LOMI 28(1972), 130-144 Engl. transl. J. Soviet Math. 23 (1975).

[2] C.M. Ringel, Tame algebras and integral quadratic forms, Lecture Notes 1099, Springer Verlag 1984.

[3] L.A. Nazarova, A.V. Roiter, Categorial matrix problems and the Brauer-Thrall conjecture, Inst. Math. Acad. Sci. Kiev (1973), German transl. Mitt. Math. Sem. Giessen 115(1975).

[4] L.A. Nazarova, A.V. Roiter, Representations of partially ordered sets, LOMI 28(1972),5-32, Engl. transl. J. Soviet Math. 23(1975).

[5] M.M. Kleiner, Partially ordered sets of finite type, LOMI 28 (1972), 32-41, Engl. transl. J. Soviet Math. 23 (1975).

[6] L.A. Nazarova, A.V. Roiter, Representations of weakly completed partially ordered sets, in Linear Algebra and Theory of Representations, Inst. Math. Acad. Sci. Kiev (1983), in russian.

[7] L.A. Nazarova, Partially ordered sets of infinite type, Izv. Akad. Nauk SSSR Ser. Mat. 39(1975), 963-991.

[8] P. Gabriel, Représentations indécomposables des ensembles ordonnés, Sem. Dubreil, Paris 1973.

[9] L.A. Nazarova, Representations of quadruples, Izv. Akad. Nauk SSSR Ser. Math. 31(1967), 1361-1377.

[10] I.M. Gelfand, V.A. Ponomarev, Problems of linear algebra and classification of quadruples of subspaces in a finite-dimensional vector space, Coll. Math. Soc. J. Bolyai, Hilbert Spaces operators, Tihany, Hungary, 1970.

[11] A.T. Zavadski, L.A. Nazarova, Partially ordered sets of finite growth, Functional Analysis and its applications 16(1982), 72-73.

[12] R. Bautista, P.Gabriel, A.V. Roiter, L. Salmerón, Representation-finite algebras and multiplicative bases, Invent. math. 81(1985), 217-185.

GEOMETRY OF REPRESENTATIONS OF QUIVERS

H. KRAFT and Ch. RIEDTMANN

Mathematisches Institut der Universitat Basel, Rheinsprung 21, CH-4051 Basel, Switzerland.

Laboratoire de Mathematiques, Institut Fourier, B.P. 74, 38402 Saint Martin-d'Heres, France.

INTRODUCTION

One of the first results about representations of quivers was Gabriel's characterization of the quivers of finite representation type and of their indecomposable representations [G1, G2] : The underlying graph of such a quiver is a union of Dynkin diagrams and the indecomposables are in one-to-one correspondence with the positive roots of the associated semi-simple Lie-algebra. Later Donovan-Freislich [DF] and independently Nazarova [N] discovered analogous relations between tame quivers and extended Dynkin diagrams. Since all remaining quivers are wild, there was little hope to get any further, except maybe in some special cases. Therefore Kac's spectacular paper [K1], where he describes the dimension types of all indecomposables of arbitrary quivers, came as a big surprise. In [K2] and [K3] Kac improved and completed his first results.

These notes are meant to be a guide to and through Kac's articles. In fact, most definitions and results are taken from his work. We reorganized them to give - we believe- a direct approach which is easy to follow. We refer to Kac's papers only for statements we do not prove completely.

Our point of view is of geometric nature -like in Kac's original work- and we use methods from algebraic geometry and transformation groups. The set of representations of a fixed dimension type is viewed as an algebraic variety on which the algebraic group of base change operators acts. In fact, it is a vectorspace with a linear group action. In this setting a number of interesting questions arise very naturally, for example the following :

What does the set of indecomposables look like ? How many components does it have, and what is the number of parameters ? What is the structure of the parameter space ? Is it always rational, and is there a (canonical) normal form ? How can one understand degenerations and deformations by means of representation theory ? What is the interpretation of the singularities in closures of isomorphism classes and of their tangent spaces ? What is the generic decomposition of the dimension type, and when is the generic representation indecomposable ? For which dimension types are there only finitely many isomorphism classes ?

Some of these questions were already answered by Kac and will be discussed in these notes too. But many of them are still open or have partial answers only in some very special cases. Furthermore it should be an important task to generalize Kac's results to quivers with relations. Again the set of representations of such a quiver of fixed dimension type forms an affine variety with the group of base change operators acting. But it will not be a vectorspace in general : it may have singularities and may even be reducible. Nevertheless the same questions as above can be asked here too, but -as far as we know- no real effort has been made yet to understand this more general situation from the geometric point of view. In particular, there is no handy description of the dimension types of the indecomposables even for finite or tame representation type. So once again there does not seem to be much hope...

We are grateful to C. Cibils, J.M. Fontaine, D. Luna and W. Messing for helpful discussions and suggestions, concerning mostly chapter 5.

1. QUIVERS AND REPRESENTATIONS

1.1. A quiver Q consists of a set Q_0 of vertices, a set Q_1 of arrows and two maps $t, h : Q_1 \to Q_0$ assigning to an arrow φ its tail $t\varphi$ and its head $h\varphi$, respectively. We do not exclude loops nor multiple arrows ; i.e., $t\varphi$ and $h\varphi$ may coincide, and $t\varphi = t\psi$, $h\varphi = h\psi$ does not imply $\varphi = \psi$. We assume that Q_0 and Q_1 are finite, and we set $Q_0 = \{1, 2, \ldots, n\}$.

Examples .

a) 

b) 

We fix an algebraically closed field k of arbitrary characteristic. A representation V of Q (over k) is a family $V(i)$, $i = 1, \ldots, n$ of finite-dimensional k-vector spaces together with a k-linear map $V(\varphi) : V(th) \to V(h\varphi)$ for each arrow φ. The vector $\underline{\dim}\, V = (\dim V(1), \ldots, \dim V(n)) \in \mathbb{N}^n$ is the dimension type of V . A morphism $f : V \to W$ is a family of k-linear maps $f(i) : V(i) \to W(i)$, $i = 1, \ldots, n$, such that $W(\varphi) \circ f(t\varphi) = f(h\varphi) \circ V(\varphi)$ for all arrows φ .

The direct sum $V \oplus W$ of two representations V and W is defined by $(V \oplus W)(i) = V(i) \oplus W(i)$ and $(V \oplus W)(\varphi) = \begin{pmatrix} V(\varphi) & 0 \\ 0 & W(\varphi) \end{pmatrix}$. A representation V is called indecomposable if $V \neq 0$ and if $V = W_1 \oplus W_2$ implies $W_1 = 0$ or $W_2 = 0$.

1.2. Q is of finite representation type if Q has only finitely many indecomposable representations, up to isomorphism. For instance, the quiver

$$Q = 1 \to 2 \to \ldots \to n-1 \to n$$

is of finite representation type ; the indecomposables are the $V_{i,j}$'s with $i \leq j$, which are defined by

$$V_{i,j}(\ell) = \begin{cases} k & \text{if } i \leq \ell \leq j \\ 0 & \text{otherwise} \end{cases}$$

$$V_{i,j}(\varphi) = \begin{cases} \mathbb{1} & \text{if } i \leq t\varphi < h\varphi \leq j \\ 0 & \text{otherwise.} \end{cases}$$

$$V_{ij} = 0 \xrightarrow{0} 0 \ldots 0 \xrightarrow{0} k \xrightarrow{1} k \ldots k \xrightarrow{1} k \xrightarrow{0} 0 \ldots \xrightarrow{0} 0 \; .$$
$$\phantom{V_{ij} = 0 \xrightarrow{0} 0 \ldots 0 \xrightarrow{0}} \underset{i}{\uparrow} \phantom{k \xrightarrow{1} k \ldots k} \underset{j}{\uparrow}$$

If there exists a full embedding of the category of representations of $\bigcirc \cdot \bigcirc$ into the category of representations of Q, Q is called <u>wild</u>. In this case, the problem of establishing a list of representatives of all indecomposables is considered hopeless. Finally, if Q is neither of finite representation type nor wild, it is said to be <u>tame</u>.

1.3. Tits form

The <u>Tits form</u> q_Q, a quadratic form on $\mathbb{Q}^n$ associated with Q, is defined as follows :

$$q(x_1,\ldots, x_n) = \sum_{i=1}^{n} x_i^2 - \sum_{\varphi \in Q_1} x_{t\varphi} x_{h\varphi} \; .$$

Obviously q_Q only depends on the non-oriented graph $\bar{Q}$ underlying Q. The <u>Cartan matrix</u> C_Q describes the bilinear form $(,)_Q$ associated with q_Q :

$$(x, y)_Q = q(x+y) - q(x) - q(y) = x \, C_Q \, y^T \; .$$

The components of $C_Q = (c_{ij})$ are :

$$c_{ij} = \begin{cases} 2 - 2 \, \#\{\text{loops in } i\} & \text{if } i = j \\ - \#\{\text{edges linking } i \text{ and } j\} & \text{if } i \neq j \; . \end{cases}$$

<u>Examples</u>.

	Q	q_Q	C_Q
a)	$\bigcirc 1 \bigcirc$	$-x_1^2$	(-2)
b)	$1 \longrightarrow 2$	$x_1^2 + x_2^2 - x_1 x_2$	$\begin{pmatrix} 2 & -1 \\ -1 & 2 \end{pmatrix}$
c)	$1 \longrightarrow 2 \bigcirc$	$x_1^2 - x_1 x_2$	$\begin{pmatrix} 2 & -1 \\ -1 & 0 \end{pmatrix}$

LEMMA [K1, lemma 1.2]. <u>Let</u> Q <u>be a connected quiver</u>, q <u>its Tits form and</u> C <u>its Cartan matrix</u>.

a) q <u>is positive definite if and only if</u> $\bar{Q}$ <u>is a Dynkin diagram</u>.

b) q <u>is positive semidefinite if and only if</u> $\bar{Q}$ <u>is an extended</u>

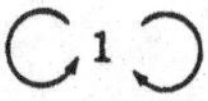

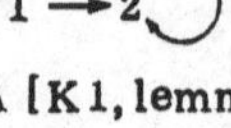

Dynkin diagram. In this case, rank C = n-1 and

$\{\alpha \in \mathbb{N}^n : C\alpha \le 0\} = \{\alpha \in \mathbb{N}^n : C\alpha = 0\} = \{\alpha \in \mathbb{N}^n : q(\alpha) = 0\} = \mathbb{N}\delta_Q$

for a unique $\delta_Q \in \mathbb{N}^n \setminus \{0\}$.

c) q is indefinite if and only if $C\alpha \ge 0$ for $\alpha \in \mathbb{N}^n$ implies $\alpha = 0$ and there exists an $\alpha \in \mathbb{N}^n$ such that $\alpha > 0$ and $C\alpha < 0$.

Here $\alpha > 0$ means that $\alpha(i) > 0$ and $\alpha \ge 0$ that $\alpha(i) \ge 0$ for all i . For a proof, see [B], [V].

Dynkin diagrams

A_m , $m \ge 1$ 1 — 2 — ... — m-1 — m

D_m , $m \ge 4$ 1 — 2 — ... — m-2 < (m-1, m)

E_m , $6 \le m \le 8$ 1 — 2 — 3 — 5 — ... — m-1 — m, with 4 joined to 3

Extended Dynkin diagrams

		δ_Q
$\tilde{A}_m$, $m \ge 1$	1 — 2 — ... — m-1 — m, with 0 joined to 1 and m	(11...11)
$\tilde{D}_m$, $m \ge 4$	(0, 1) > 2 — 3 — ... — m-2 < (m-1, m)	(1122...2211)
$\tilde{E}_6$	1 — 2 — 3 — 5 — 6, with 3 — 4 — 0	(1123221)
$\tilde{E}_7$	0 — 1 — 2 — 3 — 5 — 6 — 7, with 4 joined to 3	(12342321)
$\tilde{E}_8$	1 — 2 — 3 — 5 —...— 8 — 0, with 4 joined to 3	(124635432)

1.4. The following two fundamental results on representations of quivers are due to Gabriel (Theorem 1) and Donovan-Freislich and Nazarova (Theorem 2). Along with their generalizations to non algebraically closed fields, they can also be found in [DR] .

Recall that for a Dynkin diagram $\overline{Q}$ the vectors $\alpha \in \mathbb{N}^n$ with

$q(\alpha) = 1$ are precisely the positive roots of the corresponding semisimple Lie algebra [B]. A similar statement holds for an extended Dynkin diagram $\bar{Q}$: the $\alpha \in \mathbb{N}^n$ with $q(\alpha) = 1$ are the positive real roots and the $\alpha \in \mathbb{N}\delta_Q \setminus \{0\}$ the positive imaginary roots of the corresponding infinite dimensional Kac-Moody algebra [K4].

THEOREM 1 [G1, G2]. A connected quiver Q is of finite representation type if and only if $\bar{Q}$ is a Dynkin diagram. The map dim induces a bijection between isomorphism classes of indecomposable representations of Q and positive roots of Q.

THEOREM 2 [DF, N]. A connected quiver Q is tame if and only if $\bar{Q}$ is an extended Dynkin diagram. For each indecomposable V, $\dim V$ is a positive real or imaginary root of Q. For each positive real root α of Q, there exists a unique indecomposable V (up to isomorphism) with $\dim V = \alpha$. There exists a cofinite subset E_Q of $\mathbb{P}^1 k$ such that, for each positive imaginary root $\lambda\delta_Q$, the isomorphism classes of indecomposables V with $\dim V = \lambda\delta_Q$ are parametrized by E_Q.

2. THE REPRESENTATION SPACE OF A QUIVER

2.1. The representation space $R(Q, \alpha)$ of Q of dimension type $\alpha = (\alpha(1), \ldots, \alpha(n)) \in \mathbb{N}^n$ is the set of representations

$$R(Q, \alpha) = \{V : V(i) = k^{\alpha(i)},\ i = 1, \ldots, n\}.$$

Since $V \in R(Q, \alpha)$ is determined by the maps $V(\varphi)$, we have

$$R(Q, \alpha) = \prod_{\varphi \in Q_1} \mathrm{Hom}_k\left(k^{\alpha(t\varphi)}, k^{\alpha(h\varphi)}\right) = \prod_{\varphi \in Q_1} M_\varphi,$$

where M_φ is the set of matrices of size $\alpha(h\varphi) \times \alpha(t\varphi)$ with entries in k. We will consider $R(Q, \alpha)$ as an affine variety.

The algebraic group

$$GL(\alpha) = \prod_{i=1}^{n} GL(\alpha(i))$$

operates linearly (and regularly) on $R(Q,\alpha)$:

$$(g\cdot V)(\varphi) = g_{h\varphi} \circ V(\varphi) \circ g_{t\varphi}^{-1}$$

for $g = (g_1,\dots,g_n) \in GL(\alpha)$. The group $GL(\alpha)$ is the group of units of the finite dimensional k-algebra $M(\alpha) = \prod_{i=1}^{n} M(\alpha(i))$, where $M(s)$ is the algebra of $s\times s$-matrices. The group k^* diagonally embedded in $GL(\alpha)$ acts trivially, and we obtain an induced operation of

$$G(\alpha) = GL(\alpha)/k^*$$

on $R(Q,\alpha)$.

Using the notion of dimension for algebraic varieties, we can reinterpret the Tits form in the following way

$$q_Q(\alpha) = \dim GL(\alpha) - \dim R(Q,\alpha) .$$

2.2. By definition, the $GL(\alpha)$-orbits in $R(Q,\alpha)$ are just the isomorphism classes of representations. The <u>stabilizer</u>

$$C_{GL(\alpha)}V = \{g \in GL(\alpha) : g\cdot V = V\}$$

is the group Aut V of units in the endomorphism ring $\operatorname{End} V \subseteq M(\alpha)$. Thus it is connected.

V is <u>indecomposable</u> if End V is local; i.e., the nilpotent endomorphisms form an ideal of codimension 1 . Equivalently, $k^* \subseteq C_{GL(\alpha)}V = \operatorname{Aut} V$ is a maximal torus, which means that every semisimple element of Aut V lies in k^* .

More generally, decomposing a representation V into indecomposables corresponds to choosing a maximal torus in Aut V . Indeed, if T is a maximal torus in Aut V , we can decompose

$$V(i) = \bigoplus_\chi V_\chi(i) \quad \text{with} \quad V_\chi(i) = \{v \in V(i) : t\cdot v = \chi(t)v \quad \text{for all} \quad t \in T\}$$

for all i , where $\chi : T \to k^*$ ranges over the characters of T . Then $V(\varphi)(V_\chi(t\varphi)) \subseteq V_\chi(h\varphi)$ for all arrows φ , and we thus obtain a decomposition $V = \bigoplus_\chi V_\chi$. Since T operates on V_χ by scalar multiplication, $k^* \subseteq \operatorname{Aut} V_\chi$ is a maximal torus, and therefore V_χ is indecomposable. Conversely, if $V = V_1 \oplus \dots \oplus V_r$ with V_i indecomposable, the product of the maximal tori $k^* \subseteq \operatorname{Aut} V_i$ is a maximal torus of Aut V .

The map $g \mapsto g \cdot V$ induces an isomorphism

$$GL(\alpha)/C_{GL(\alpha)}V \to \mathcal{O}_V ,$$

where $\mathcal{O}_V$ is the orbit of V. This implies

$$\dim \mathcal{O}_V + \dim \operatorname{End} V = \dim GL(\alpha) .$$

Since $k \subseteq \operatorname{End} V$ for any representation V, we get

$$\dim \mathcal{O}_V \leq \dim GL(\alpha) - 1 .$$

Using this inequality, Tits found a very nice argument, which proves part of theorem 1 in 1.4 [G2]. Assume that Q is a connected quiver of finite representative type and choose $\alpha \in \mathbb{N}^n \backslash \{0\}$. Since any representation of Q can be decomposed into a direct sum of indecomposables, $R(Q, \alpha)$ contains only finitely many $GL(\alpha)$-orbits. So one orbit must be dense and thus have the same dimension as $R(Q, \alpha)$. Therefore

$$\dim R(Q, \alpha) \leq \dim GL(\alpha) - 1$$

or equivalently

$$q_Q(\alpha) \geq 1 .$$

Since all off-diagonal entries of C_Q are non-positive, it follows that q_Q is positive definite on $\mathbb{Z}^n$ and hence on $\mathbb{Q}^n$. Thus $\bar{Q}$ is a Dynkin diagram (lemma 1.3).

2.3. In this paragraph we study $R(Q, \alpha)$, its decomposition into sheets (2.4) and the indecomposables in each sheet for a particular example, which should serve as motivation and illustration for the general definitions. The notations used here are adapted to those introduced later.

We consider the wild quiver

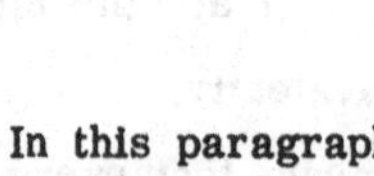

and the dimension vector $\alpha = (2, 1)$. We have

$$\dim R(Q, \alpha) = 6 ,$$
$$\dim GL(\alpha) = 5 ,$$
$$q_Q(x, y) = x^2 + y^2 - 3xy \quad , \quad q_Q(\alpha) = -1 ,$$
$$(\alpha, (10))_Q = 1 \quad , \quad (\alpha, (01))_Q = -4 .$$

The set $C = \{V : \det\binom{V(\varphi)}{V(\psi)} = 0\}$ is closed and irreducible in $X = R(Q,\alpha)$. Every representation V in $X^{(1)} = X \setminus C$ has a unique representative of the form

$$V(\varphi) = (1\ 0)\ ,\quad V(\psi) = (0\ 1)\ ,\quad V(\chi) = \binom{\alpha}{\beta} \in k^2\ .$$

Note that $X^{(1)}$ is irreducible. It consists of indecomposable representations with endomorphism ring k.

Every representation in C is isomorphic to precisely one of the following

i) $V(\varphi) = (\alpha\ 0)\ ,\quad V(\psi) = (\beta\ 0)\ ,\quad V(\chi) = \binom{0}{1}$ with $(\alpha\ \beta) \in \mathbb{P}^1 k$,

ii) $V(\varphi) = (\alpha\ 0)\ ,\quad V(\psi) = (\beta\ 0)\ ,\quad V(\chi) = \binom{1}{0}$ with $(\alpha\ \beta) \in k^2 \setminus \{0\}$,

iii) $V(\varphi) = (\alpha\ 0)\ ,\quad V(\psi) = (\beta\ 0)\ ,\quad V(\chi) = \binom{0}{0}$ with $(\alpha\ \beta) \in \mathbb{P}^1 k$,

iv) $V(\varphi) = V(\psi) = (0\ 0)\ ,\quad V(\chi) = \binom{1}{0}$,

v) $V(\varphi) = V(\psi) = (0\ 0)\ ,\quad V(\chi) = \binom{0}{0}$.

The representations of type i) are indecomposable with endomorphism ring $k[t]/(t^2)$, all others are decomposable with endomorphism rings of dimension 2, 3, 3 and 5 for the types ii), iii), iv) and v), respectively. Denote by $X^{(d)}$ the set of representations with d-dimensional endomorphism ring.

These sets can be described as follows :

$$X^{(2)} = \{V \in X : \operatorname{rank}\binom{V(\varphi)}{V(\psi)} = 1\ ,\ V(\chi) \neq 0\}\ ,$$

$$X^{(3)} = \{V \in X : \operatorname{rank}\binom{V(\varphi)}{V(\psi)} = 1\ ,\ V(\chi) = 0\} \cup$$

$$\{V \in X : V(\varphi) = V(\psi) = 0\ ,\ V(\chi) \neq 0\}\ ,$$

$$X^{(5)} = \{0\}\ .$$

$X^{(3)}$ has two disjoint irreducible components of dimension 2 and 3, respectively, whereas $X^{(2)}$ and $X^{(5)}$ are irreducible. The set of indecomposables within each $X^{(d)}$ is closed. Indeed, $V \in X^{(2)}$ is indecomposable if and only if $V(\varphi) \circ V(\chi) = 0 = V(\psi) \circ V(\chi)$. But the set of all indecomposables is neither open nor closed nor locally closed (= open $\cap$ closed) in X. We will see in 2.5 that these are general facts.

Remark. This example shows that $R(Q,\alpha)$ may contain a dense open set of indecomposables without the set of all indecomposables being open. This contradicts the statements in [K1, 2.8], [K2, §4] which lead to the definition of the "canonical decomposition of α" [K1, (2.24)].

2.4. We introduce some notions and results used later for the general setting of an algebraic group G operating regularly on an irreducible variety Z (cf. [Kr 2, II.2)].

For any $z \in Z$, the orbit $G\cdot z$ is open in $\overline{G\cdot z}$. In particular, if $z' \in \overline{G\cdot z}\setminus G\cdot z$, then $\dim G\cdot z' < \dim G\cdot z$, or equivalently $\dim C_G z' > \dim C_G z$, where $C_G z = \{g \in G : g\cdot z = z\}$.

The fixed point set $Z^g = \{z \in Z : g\cdot z = z\}$ is closed in Z for any $g \in G$: identify Z^g with the inverse image of the diagonal under the regular map $Z \to Z\times Z$ given by $z \mapsto (z, g\cdot z)$. Thus

$$Z^G = \{z \in Z : g\cdot z = z, \ \forall g \in G\}$$

is closed as well.

For $s \in \mathbb{N}$ the set

$$Z_{(s)} = \{z \in Z : \dim G\cdot z = s\}$$

is locally closed in Z, since by Chevalley's theorem ([EGA IV, §13], cf. [Kr 2, II.2.6]) the function $z \mapsto \dim C_G z$ is upper semicontinuous. In particular, the union Z^{max} of all orbits of maximal dimension is open and dense in Z. An irreducible component $\mathcal{S}$ of a $Z_{(s)}$ is called a <u>sheet</u> of Z for the action of G. All orbits in $\mathcal{S}$ are closed in $\mathcal{S}$ and have the same dimension.

As an example, we consider the operation of $G = GL(n)$ on $Z = M(n)$ by conjugation. With a matrix A having eigenvalues $\lambda_1, \dots, \lambda_r$ and Jordan blocks with eigenvalue λ_i of size $\rho_{i1} \ge \rho_{i2} \ge \dots \ge \rho_{in} \ge 0$ for $i = 1, \dots, r$, we associate the partition $p_A = (p_1, \dots, p_n)$ of n, where $p_j = \sum_{i=1}^{r} \rho_{ij}$. This is the partition corresponding to the dimensions of the invariant factor modules for the n-dimensional $k[T]$-module given by A. It is easy to see that all matrices A yielding the same partition belong to the same sheet of Z. In fact, we have the following result, which is due to Dixmier, Peterson, Kraft (cf. [Kr 1], [P]).

PROPOSITION. The map $A \mapsto p_A$ induces a bijection between sheets of $M(n)$ and partitions of n. The sheets are disjoint. They are smooth, and each one contains exactly one nilpotent conjugacy class and a dense open set of semisimple matrices. The orbit space $\mathcal{S}_p/GL(n)$ of the sheet corresponding to $p = (p_1, \dots, p_n)$ is isomorphic to k^{p_1}.

2.5. Fix Q and set $R(\alpha) = R(Q, \alpha)$. Put

$$R(\alpha)^{(d)} = \{V \in R(\alpha) : \dim \operatorname{End} V = d\}$$

for $d \in \mathbb{N}$.

PROPOSITION. a) $R(\alpha)^{(d)}$ is locally closed in $R(\alpha)$; $R(\alpha)^{max}$ is open and dense in $R(\alpha)$.

b) $R(\alpha)^{(d)}_{ind} = \{V \in R(\alpha)^{(d)} : V \text{ indecomposable}\}$ is closed in $R(\alpha)^{(d)}$.

As a consequence, $R(\alpha)_{ind} = \bigcup_{d \in \mathbb{N}} R(\alpha)^{(d)}_{ind}$ is a constructible set; i.e., a finite union of locally closed sets.

Proof. a) follows from 2.4, since

$$R(\alpha)^{(d)} = R(\alpha)_{\bar{d}}$$

with $\bar{d} = \dim GL(\alpha) - d$.

For b) consider the closed subvariety

$$N = \{(V, \rho) \in R(\alpha) \times M(\alpha) : \rho \in \operatorname{End} V,\ \rho \text{ nilpotent}\}$$

and the projection

$$p : N \to R(\alpha).$$

The fiber $p^{-1}(V)$ of a representation $V \in R(\alpha)$ consists of the nilpotent endomorphisms of V. Since the zero section $R(\alpha) \to N$ meets every irreducible component of every fiber, the function $V \mapsto \dim p^{-1}(V)$ is upper semicontinuous (theorem of Chevalley). But $V \in R(\alpha)$ is indecomposable if and only if $\dim p^{-1}(V) \geq \dim \operatorname{End} V - 1$, and so

$$R(\alpha)^{(d)}_{ind} = \{V \in R(\alpha)^{(d)} : \dim p^{-1}(V) \geq d-1\}$$

is closed in $R(\alpha)^{(d)}$.

<u>Remark</u> (Happel). If $V = V' \oplus V'' \in R(\alpha)^{max}$, then $Ext^1(V', V'') = 0$. Indeed, if there exists a non-split extension

$$0 \to V'' \to W \to V' \to 0 ,$$

Then $\mathcal{O}_V \subseteq \bar{\mathcal{O}}_W \setminus \mathcal{O}_W$, which contradicts $V \in R(\alpha)^{max}$.

We embed $GL(\alpha)$ as blocks along the diagonal into $M(N)$, $N = \sum_{i=1}^{n} \alpha(i)$. A <u>minisheet</u> of $GL(\alpha)$ is an irreducible component of the intersection $\mathcal{S} \cap GL(\alpha)$, where $\mathcal{S}$ is a sheet of $M(N)$ with respect to the operation of $GL(N)$ by conjugation (2.4). Part a) of the following lemma implies that each minisheet is contained in a sheet of $GL(\alpha)$, where we consider the action of $GL(\alpha)$ on itself by conjugation.

LEMMA. a) <u>The functions</u> $g \mapsto \dim C_{GL(\alpha)}g$ <u>and</u> $g \mapsto \dim R(\alpha)^g$ <u>are constant on minisheets</u>.

b) <u>Each minisheet contains a dense set of semisimple elements</u>.

<u>Proof</u>. a) $GL(\alpha)$ operates on $Hom_k(k^{\alpha(i)}, k^{\alpha(j)})$ by $g \circ f = g_j \circ f \circ g_i^{-1}$ for $g = (g_1, \ldots, g_n)$, and the function $g \mapsto \dim Hom(k^{\alpha(i)}, k^{\alpha(j)})^g$ is upper semicontinuous on $GL(\alpha)$. On the other hand, we have

$$M(N) = \prod_{i,j} Hom(k^{\alpha(i)}, k^{\alpha(j)})$$

and

$$M(N)^g = \prod_{i,j} Hom(k^{\alpha(i)}, k^{\alpha(j)})^g$$

for $g \in GL(\alpha)$. The function $g \mapsto \dim M(N)^g$ is constant on each sheet $\mathcal{S}$ of $M(N)$ and hence constant on minisheets. Therefore the functions $g \mapsto \dim Hom(k^{\alpha(i)}, k^{\alpha(j)})^g$ are also constant on minisheets. But we have

$$R(\alpha)^g = \prod_{\varphi \in Q_1} Hom(k^{\alpha(t\varphi)}, k^{\alpha(h\varphi)})^g$$

and

$$C_{GL(\alpha)}g = \{\text{units of } \prod_{i=1}^{n} (End\ k^{\alpha(i)})^g\}$$

for $g \in GL(\alpha)$.

b) Let $\mathcal{S}$ be a sheet of $M(N)$ and $\mathcal{S}'$ an irreducible component of $\mathcal{S} \cap GL(\alpha)$. Choose an element $x \in \mathcal{S}'$ which does not lie in any other irreducible component of $\mathcal{S} \cap GL(\alpha)$. Considering the component of x in each $GL(\alpha(i))$ separately, we may suppose that x is in Jordan normal form. As an easy consequence of the description of sheets in $M(N)$ given in 2.4, we find an invertible diagonal matrix $d \in \mathcal{S}$ such that the line

$$L = \{\lambda s + (1-\lambda)d : \lambda \in k\}$$

is contained in $\mathcal{S}$. Hence $L' = L \cap GL(\alpha)$ is an irreducible curve in $\mathcal{S} \cap GL(\alpha)$ containing x and d. By the choice of x, L' -and thus d- is contained in $\mathcal{S}'$.

So we found one semisimple element in $\mathcal{S}'$. But the set of semisimple elements in $\mathcal{S}$ is open and dense (2.4), and therefore $\mathcal{S}'$ contains a dense set of semisimples.

2.6. DEFINITION. $V \in R(\alpha)$ is <u>stably indecomposable</u> if there exists an open neighborhood of V consisting of indecomposable representations.

THEOREM. V <u>is stably indecomposable if and only if</u> $\operatorname{End} V = k$.

<u>Proof.</u> If $\operatorname{End} V = k$, all representations in the dense open set $R(\alpha)^{max}$ have endomorphism ring k.

Conversely, suppose that V is indecomposable and has an automorphism $g_0 \notin k^*$. Choose an open neighborhood U of V. We want to show that U contains a representation admitting a semisimple automorphism outside of k^*. Then V cannot be stably indecomposable. Set

$$S = \{g \in GL(\alpha) : \dim R(\alpha)^g = \dim R(\alpha)^{g_0}\}$$

and

$$E = \bigcup_{W \in U} \operatorname{End} W \subseteq M(\alpha) .$$

Since $g_0 \notin k^*$, S does not intersect k^*. Moreover, S contains a dense set of semisimple elements, since it is a union of minisheets (2.5). The following lemma implies that $E \cap S$ is open in S. Hence E contains semisimple elements, and the theorem follows.

LEMMA. <u>Let</u> G <u>be an algebraic group operating linearly on a vectorspace</u> V, $U \subseteq V$ <u>an open subset</u>, $g_0 \in G$. <u>Set</u>
$S = \{g \in G : \dim V^g = \dim V^{g_0}\}$.
<u>Then</u>

$S' = \{ g \in S : \exists u \in U \text{ with } g \cdot u = u \}$

<u>is open in</u> S.

<u>Proof</u>. Consider the vector bundle

$p_1 : S \times V \to S$.

By the definition of S,

$L = \{ (s, v) : s \cdot v = v \}$

is a subbundle. Since the restriction $\varphi = p_1/L : L \to S$ is flat, the image

$S' = \varphi(L \cap S \times U)$

is open.

2.7. <u>Generic decomposition.</u>

PROPOSITION. <u>For</u> $\alpha \in \mathbb{N}^n$ <u>there exists a unique decomposition</u> $\alpha = \alpha_1 + \ldots + \alpha_s$ <u>such that the set</u>

$R(\alpha)_{gen} = \{ V \in R(\alpha) : V = V_1 \oplus \ldots \oplus V_s , \ \underline{\dim}\, V_i = \alpha_i , \ V_i \text{ indecomposable} \}$

<u>contains an open and dense set of</u> $R(\alpha)$.

$\alpha = \alpha_1 + \ldots + \alpha_s$ is called the <u>generic decomposition</u> of α, representations in $R(\alpha)^{max}_{gen} = R(\alpha)_{gen} \cap R(\alpha)^{max}$ are called <u>generic representations of type</u> α.

<u>Remarks</u>.

a) As example 2.3 shows, $R(\alpha)_{gen}$ is not necessarily open in $R(\alpha)$.

b) The generic decomposition depends on the orientation of Q: choose $\bar{Q} =$  and $\alpha = \begin{smallmatrix} & 2 & \\ 1 & & 1 \end{smallmatrix}$. For the orientation , the generic decomposition is $\begin{smallmatrix} & 1 & \\ 1 & & 1 \end{smallmatrix} + \begin{smallmatrix} & 1 & \\ 0 & & 0 \end{smallmatrix}$, whereas α is generically indecomposable for the orientation .

<u>Proof</u>. For each decomposition $\alpha = \beta_1 + \ldots + \beta_t$, $\beta_i \in \mathbb{N}^n$, the set of representations V such that $V = V_1 \oplus \ldots \oplus V_t$ with $\underline{\dim}\, V_i = \beta_i$ is constructible, since it can be viewed as the image of $GL(\alpha) \times R(\beta_1) \times \ldots \times R(\beta_t)$ under the map $(g, V_1, \ldots, V_t) \mapsto g \cdot (V_1 \oplus \ldots \oplus V_t)$. Thus the set

$$R(\alpha;\beta_1,\ldots,\beta_t) = \{V : V = V_1\oplus\ldots\oplus V_t,\ \underline{\dim}\ V_i = \beta_i\ ,\ V_i \text{ indecomposable}\}$$

is constructible as well, and $R(\alpha)$ is the disjoint union of the $R(\alpha\,;\,\beta_1,\ldots,\beta_t)$, taken over the finite set of all distinct decompositions of α. So precisely one of these sets, say $R(\alpha\,;\,\alpha_1,\ldots,\alpha_s)$, contains an open dense set of $R(\alpha)$.

As a consequence of this and theorem 2.6 we obtain

COROLLARY. Let $V = V_1\oplus\ldots\oplus V_s$ be generic, V_i indecomposable. Then $\mathrm{End}\,V_i = k$, for all i, and $\mathrm{Ext}^1(V_i,V_j) = 0$ for $i \neq j$. In particular, if the generic representation V is indecomposable, we have $R(\alpha)^{\max} \subseteq R(\alpha)_{\mathrm{ind}}$ and $\mathrm{End}\,V = k$.

3. THE FUNDAMENTAL SET

3.1. Denote by $\epsilon_1,\ldots,\epsilon_n$ the standard basis of $\mathbb{Q}^n$.

The fundamental set F_Q is defined by

$$F_Q = \{\alpha \in \mathbb{N}^n\setminus\{0\} : (\alpha,\epsilon_i) \leq 0\ ,\ \operatorname{supp}\alpha \text{ connected}\}.$$

Here $(\ ,\ )$ is the bilinear form $(\ ,\ )_Q$ defined in 1.3, and $\operatorname{supp}\alpha$ denotes the full subquiver of Q whose vertices are $\{i : \alpha(i)\neq 0\}$.

The following result is an easy consequence of lemma 1.3.

LEMMA 1. Let Q be connected.

a) $F_Q = \emptyset$ if and only if $\bar{Q}$ is a Dynkin diagram.

b) $F_Q = \mathbb{N}\delta\setminus\{0\}$ for some $\delta \neq 0$ if and only if $\bar{Q}$ is an extended Dynkin diagram ; in this case $\delta = \delta_Q$.

c) If $q_Q(\alpha) = 0$ for some $\alpha \in F_Q$, then $\operatorname{supp}\alpha$ is tame (i.e. $\overline{\operatorname{supp}\alpha}$ is an extended Dynkin diagram).

LEMMA 2. Let $\alpha = \beta_1+\ldots+\beta_r \in F_Q$ with $r \geq 2$ and $\beta_1,\ldots,\beta_r \in \mathbb{N}^n\setminus\{0\}$ such that $q(\alpha) \geq q(\beta_1)+\ldots+q(\beta_r)$. Then $\operatorname{supp}\alpha$ is tame, and α is a multiple of $\delta_{\operatorname{supp}\alpha}$.

Proof. We first consider the case $r = 2$ and set $\beta_1 = \gamma = \Sigma c_i\epsilon_i$, $\beta_2 = \delta = \Sigma d_i\epsilon_i$, $\alpha = \Sigma a_i\epsilon_i$. We may suppose $Q = \operatorname{supp}\alpha$. By assumption we have $(\gamma, \delta) = q(\alpha) - q(\gamma) - q(\delta) \geq 0$. An easy computation based on $a_i = c_i + d_i$ and $c_{ij} = c_{ji}$ for the coefficients of the Cartan matrix $C = (c_{ij})$ of Q (1.3) yields

$$0 \leq (\gamma, \delta) = \sum_{i,j} c_{ij} c_i d_j$$

$$= \sum_j \frac{c_j d_j}{a_j} \sum_i c_{ij} a_i + \tfrac{1}{2} \sum_{i \neq j} c_{ij} \left(\frac{c_i}{a_i} - \frac{c_j}{a_j}\right)^2 a_i a_j .$$

Since

$$(\alpha, \varepsilon_j) = \sum_i c_{ij} a_i \leq 0 \text{ for all } j \text{ and } c_{ij} \leq 0 \text{ for all } i \neq j ,$$

this inequality implies

$$\frac{c_i}{a_i} = \frac{c_j}{a_j} \text{ if } c_{ij} \neq 0 .$$

But Q is connected, and therefore α and γ are proportional. As a consequence, we have $(\alpha, \varepsilon_j) = 0$ for all j , hence $C\alpha = 0$. But then Q is tame and $\alpha \in \mathbb{N}\delta_Q$ (lemma 1.3c)).

In case $r > 2$, we have

$$(\alpha, \alpha) = \sum_i (\alpha, \beta_i) \geq \sum_i (\beta_i, \beta_i) .$$

This implies

$$(\alpha - \beta_i, \beta_i) \geq 0$$

for some i , and we apply what we already proved to $\gamma = \beta_i$, $\delta = \alpha - \beta_i$.

3.3. THEOREM. <u>If</u> α <u>lies in</u> F_Q <u>and</u> supp α <u>is not tame, then the generic representation in</u> $R(\alpha)$ <u>is indecomposable</u>.

<u>Proof</u>. Let $\alpha = \alpha_1 + \ldots + \alpha_s$ be the generic decomposition, and suppose $s \geq 2$. Set

$$R' = R(\alpha_1) \times \ldots \times R(\alpha_s)$$

and

$$G' = GL(\alpha_1) \times \ldots \times GL(\alpha_s) .$$

The image of

$$\varphi : GL(\alpha) \times R' \longrightarrow R(\alpha)$$
$$(g, V) \longmapsto g \cdot V$$

is dense in $R(\alpha)$ by construction, and φ is constant on the orbits of the free action of G' on $GL(\alpha) \times R'$ given by $h \cdot (g, V) = (gh^{-1}, h \cdot V)$. As a consequence,

$$\dim\ GL(\alpha) + \dim R' - \dim G' \geq \dim R(\alpha) ,$$

which implies

$$q(\alpha) \geq q(\alpha_1) + \ldots + q(\alpha_s) ,$$

in contradiction with lemma 2 of 3.2.

3.4. Number of parameters.

Let G be an algebraic group acting on a variety Z. If $X \subseteq Z$ is a G-stable subset, we write

$$X = \cup X_{(s)}$$

with

$$X_{(s)} = \{x \in X : \dim \mathcal{O}_x = s\} .$$

DEFINITION. *The number of parameters of* X, *is*

$$\mu(X) = \max_s (\dim X_{(s)} - s) .$$

Here $\dim X_{(s)}$ denotes the dimension of the closure of $X_{(s)}$ in Z.

Example. If the generic representation in $R(\alpha)$ is indecomposable, we have

$$\mu(R(\alpha)^{max}) = \dim R(\alpha) - (\dim\ GL(\alpha) - 1) = 1 - q(\alpha)$$

(corollary 2.7).

THEOREM. *If* α *lies in* F_Q *and* supp α *is not tame, then*

$$\mu(R(\alpha)_{ind}) = \mu(R(\alpha)^{max}) = 1 - q(\alpha) > \mu(R(\alpha)_{ind}^{(d)})$$

for all $d > 1$.

For the proof we need :

PROPOSITION. *Let* G *act on* Z, *and suppose that* G *contains a finite number of unipotent conjugacy classes. Then the number of parameters of*

$$X = \{z \in Z : G_z \text{ unipotent}\}$$

satisfies

$$\mu(X) \leq \max_{\substack{u \in G \\ u \text{ unipotent}}} (\dim Z^u - \dim G_u) .$$

Remark. The Jordan normal form shows that $GL(\alpha)$ and also $G(\alpha)$ contain only finitely many unipotent conjugacy classes. In fact, this holds for any reductive group [L].

Proof. Consider the closed subvariety

$$L = \{(g, z) \in G \times Z \; : \; g \cdot z = z\}$$

of $G \times Z$ and the projections

$$\varphi = pr_2 : L \to Z \; , \quad \psi = pr_1 : L \to G \; .$$

For $z \in Z$, we have

$$\varphi^{-1}(z) = C_G z \times \{z\} \; .$$

If z lies in $X_{(s)}$, $\dim C_G z = \dim G - s$, and therefore

$$\dim \varphi^{-1}(X_{(s)}) = \dim X_{(s)} + \dim G - s \; .$$

Consequently,

$$\mu(X) = \max_s (\dim X_{(s)} - s) = -\dim G + \max_s \dim \varphi^{-1}(X_{(s)})$$
$$= -\dim G + \dim \varphi^{-1}(X) \; .$$

The definition of X implies $\varphi^{-1}(X) \subseteq \psi^{-1}(U)$, where

$$U = \{u \in G \; : \; u \text{ unipotent}\} \; .$$

Since U consists of a finite number of conjugacy classes

$$C_u = \{gug^{-1} \; : \; g \in G\},$$

we obtain

$$\mu(X) \leq -\dim G + \max_u \dim \psi^{-1}(C_u) \; .$$

But

$$\psi^{-1}(g) = \{g\} \times Z^g \quad \text{for } g \in G,$$

and thus

$$\dim \psi^{-1}(C_u) = \dim Z^u + \dim C_u = \dim Z^u + \dim G - \dim G_u \; .$$

This proves the proposition.

Proof of the theorem. Recall that $V \in R(\alpha)$ is indecomposable if and only if $C_{G(\alpha)}V$ is unipotent, where $G(\alpha) = GL(\alpha)/k^*$ (2.1). By 3.3, $R(\alpha)^{max}$ is contained in $R(\alpha)_{ind}$, and we already saw that $\mu(R(\alpha)^{max}) = 1 - q(\alpha)$. Set $\bar{R} = R(\alpha) \setminus R(\alpha)^{max}$. The lemma below implies that

$$\dim \bar{R}^u - \dim G(\alpha)_u = \dim \bar{R}^u - \dim GL(\alpha)_u + 1 < 1 - q(\alpha)$$

for any unipotent element $u \neq \mathbb{1}$. For $u = \mathbb{1}$,

$$\dim \bar{R} - \dim G(\alpha) < \dim R - \dim GL(\alpha) + 1 < 1 - q(\alpha) \; .$$

Applying the proposition to $\bar{R}$ and $G(\alpha)$, we find

$$\mu(\bar{R}_{ind}) = \max_{d>1} \mu(R(\alpha)^{(d)}_{ind}) < 1-q(\alpha) .$$

LEMMA. <u>If</u> α <u>belongs to</u> F_Q <u>and</u> supp α <u>is not tame, then</u>
$\dim GL(\alpha)_g - \dim R(\alpha)^g > q(\alpha)$
<u>for</u> $g \in GL(\alpha)\setminus k^*$.

<u>Proof.</u> The left hand side being constant on minisheets (2.5), we may suppose that g is semisimple. Let $\alpha = \alpha_1+...+\alpha_s$ be the decomposition obtained from the eigen space decomposition of g , and note that $s \geq 2$ since $g \notin k^*$. Then we have

$$GL(\alpha)_g = \prod GL(\alpha_i) \quad \text{and} \quad R(\alpha)^g = \prod R(\alpha_i)$$

and consequently (3.2)

$$\dim GL(\alpha)_g - \dim R(\alpha)^g = \sum_{i=1}^{s} q(\alpha_i) > q(\alpha) .$$

<u>Remark.</u> The theorem shows that for $\alpha \in F_Q$ with supp α not tame, the number of parameters of indecomposables in the maximal sheet is strictly bigger than in all other sheets. In fact, the proof given in chapter 5 that this is true whenever the generic representation is indecomposable. It is an open question whether the maximal number of parameters of indecomposables always occurs in $R(\alpha)^{(d)}_{ind}$ for the smallest number d with $R(\alpha)^{(d)}_{ind} \neq \phi$.

3.5. <u>Remark.</u> From the classification of indecomposables for extended Dynkin diagrams (cf. [DR]) one obtains : <u>if</u> Q <u>is tame,</u> $R(\alpha)_{ind}$ <u>is contained in</u> $R(\alpha)^{max}$. <u>For</u> $\alpha = \lambda\delta_Q$, <u>the number of parameters is</u> λ <u>for</u> $R(\alpha)^{max}$ <u>and</u> $1 = 1-q(\alpha)$ <u>for</u> $R(\alpha)_{ind}$; <u>the generic decomposition is</u> $\alpha = \delta_Q+...+\delta_Q$.

<u>Examples</u>.

a) $Q = 1 \rightrightarrows 2$, $\alpha = (2,2)$.

$$R(\alpha)^{max} = GL(\alpha)\cdot \left\{V : V(\varphi) = \mathbb{1} ,\ V(\psi) = \begin{pmatrix}\lambda_1 & 0\\ 0 & \lambda_2\end{pmatrix} ,\ \lambda_1 \neq \lambda_2 \in k\right.$$

$$\left. \text{or } V(\psi) = \begin{pmatrix}\lambda & 1\\ 0 & \lambda\end{pmatrix} ,\ \lambda \in k \right\} \cup$$

$$GL(\alpha)\bullet \left\{V : V(\psi) = \mathbb{1}\ ,\ V(\varphi) = \begin{pmatrix}\lambda_1 & 0\\ 0 & \lambda_2\end{pmatrix},\ \lambda_1 \neq \lambda_2 \in k \right.$$
$$\left. \text{or } V(\varphi) = \begin{pmatrix}\lambda & 1\\ 0 & \lambda\end{pmatrix},\ \lambda \in k \right\}.$$

$$R(\alpha)^{ind} = GL(\alpha)\bullet \left\{V : V(\varphi) = \mathbb{1}\ ,\ V(\psi) = \begin{pmatrix}\lambda & 1\\ 0 & \lambda\end{pmatrix},\ \lambda \in k\right\} \cup$$
$$GL(\alpha)\bullet \left\{V : V(\varphi) = \begin{pmatrix}\lambda & 1\\ 0 & \lambda\end{pmatrix},\ \lambda \in k\ ,\ V(\psi) = \mathbb{1}\right\}.$$

b) $Q = 1 \xrightarrow{\varphi} 2 \xrightarrow{\psi} 3$, $1 \xrightarrow{\chi} 3$, $\alpha = (1, 2, 1)$.

$$R(\alpha)_{ind} = GL(\alpha)\cdot S \quad \text{with} \quad S(\varphi) = \begin{pmatrix}1\\0\end{pmatrix},\ S(\psi) = (0\ 1)\ ,\ S(\chi) = 1\ ,$$
$$R(\alpha)^{max} = GL(\alpha)\cdot\left\{V : V(\varphi) = \begin{pmatrix}1\\0\end{pmatrix},\ V(\psi) = (1\ 0)\ ,\ V(\chi) = \lambda \in k\right\}$$
$$\dot{\cup}\ R(\alpha)_{ind}\ .$$

4. INDECOMPOSABLES AND ROOT SYSTEMS

4.1. Reflection functors

Let Q and α be as before. Fix a source i of Q, and suppose

$$\sum_{t\varphi=i} \alpha(h\varphi) \geq \alpha(i)\ .$$

Consider the set

$$R'(Q,\alpha) = \left\{V \in R(Q,\alpha)\ :\ [V(\varphi)] : V(i) \longrightarrow \bigoplus_{t\varphi=i} V(h\varphi)\right\} \text{ injective }.$$

Obviously $R(Q,\alpha)_{ind}$ is contained in $R'(Q,\alpha)$, and $R(\alpha)_{ind} = \phi$ if α does not satisfy the required inequality, unless $\alpha = \epsilon_i$.

The quiver Q^* is obtained from Q by reversing all arrows with tail i , and α^* is given by

$$\alpha^*(k) = \begin{cases} \alpha(k) & \text{for } k \neq i \\ \sum_{t\varphi=i} \alpha(h\varphi) - \alpha(i) & \text{for } k = i\ . \end{cases}$$

We have

$$\sum_{h\varphi=i} \alpha^*(t\varphi) \geq \alpha^*(i)\ .$$

Example.

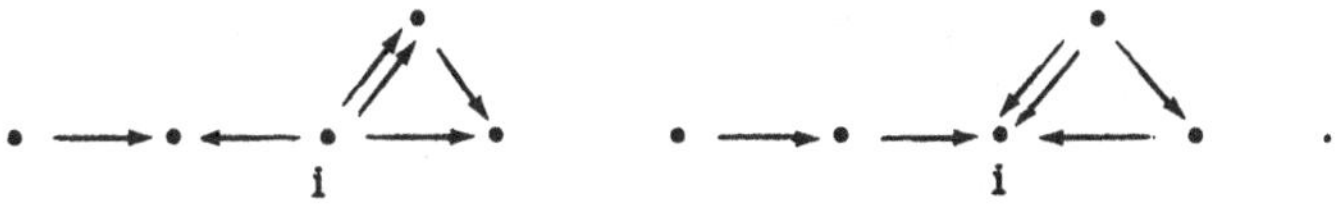

Q Q*

We set

$$R'(Q^*,\alpha^*) = \{V \in R(Q^*,\alpha^*) : [V(\varphi)] : \bigoplus_{h\varphi=i} V(t\varphi) \to V(i) \text{ surjective}\} .$$

PROPOSITION. <u>There exists a homeomorphism</u>

$$R'(Q,\alpha)/GL(\alpha) \xrightarrow{\sim} R'(Q^*,\alpha^*)/GL(\alpha^*)$$

<u>such that corresponding representations have isomorphic endomorphism rings.</u>

(We use the quotient topology of the Zariski topology).

<u>Proof</u>. Set $m = \sum_{\substack{\varphi \in Q_1 \\ t\varphi = i}} \alpha(h\varphi)$, $W = k^m$, and

$$\bar{R} = \prod_{\substack{\varphi \in Q_1 \\ t\varphi \neq i}} \mathrm{Hom}(k^{\alpha(t\varphi)}, k^{\alpha(h\varphi)}) , \quad \bar{G} = \prod_{j \neq i} GL(\alpha(j)) .$$

The required homeomorphism is obtained from the following diagram, in which we use the isomorphisms

$$R'(Q,\alpha)/GL(\alpha(i)) \xrightarrow{\sim} \bar{R} \times Gr_{\alpha(i)} W$$

$$V \longmapsto \left((V(\varphi))_{t\varphi \neq i}, \mathrm{im}[V(\varphi)]_{t\varphi = i}\right)$$

$$R'(Q^*,\alpha^*)/GL(\alpha^*(i)) \xrightarrow{\sim} \bar{R} \times Gr_{\alpha(i)} W$$

$$V \longmapsto \left((V(\varphi))_{h\varphi \neq i}, \ker[V(\varphi)]_{h\varphi = i}\right) .$$

Here $Gr_{\alpha(i)} W$ denotes the Grassmann variety of $\alpha(i)$-dimensional subspaces of W .

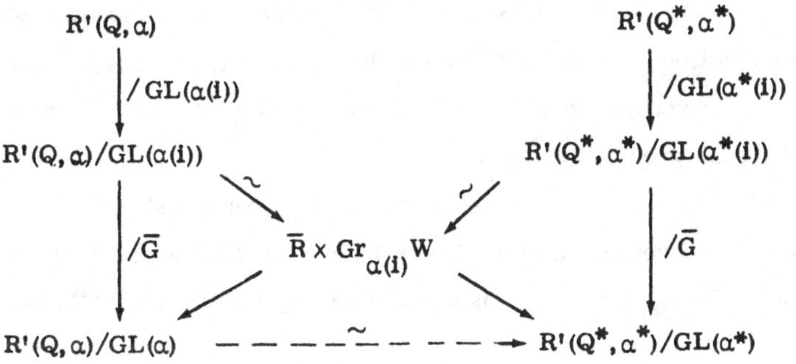

The claim about endomorphism rings follows from

$$C_{\overline{G}\times GL(\alpha(i))}V \xrightarrow{\sim} C_{\overline{G}}\overline{V} ,$$

$$C_{\overline{G}\times GL(\alpha^*(i))}V^* \xrightarrow{\sim} C_{\overline{G}}\overline{V}^*$$

for $V \in R(Q,\alpha)$, $V^* \in R(Q^*,\alpha^*)$, where $\overline{V}$ and $\overline{V}^*$ are the images in $\overline{R}\times Gr_{\alpha(i)}W$.

COROLLARY. The number of parameters as well as the number of irreducible components of maximal dimension coincide for $R(Q,\alpha)^{(d)}_{ind}$ and $R(Q^*,\alpha^*)^{(d)}_{ind}$, for all $d \in \mathbb{N}$.

Remarks. The isomorphism above is induced from the "reflection functor" of Bernstein-Gel'fand-Ponomarev [BGP] , which plays a crucial role in the proof of the two theorems of chapter 1. Independently it was introduced by Sato and Kimura under the name of "Castling transform" [SK] .

We could have started from a sink instead of a source, considering $i \in Q_0^*$ first. An admissible vertex is a source or a sink. In particular, no loop is attached at an admissible vertex. We will say that (Q^*,α^*) is obtained from (Q,α) by applying the "reflection" R_i at the admissible vertex i of Q .

It follows from the preceding proposition that all the results we proved in chapter 3 for $\alpha \in F_Q$ still hold for representations of a quiver $\widetilde{Q}$ of dimension type $\widetilde{\alpha}$, provided that $(\widetilde{Q},\widetilde{\alpha})$ is obtained from (Q,α) by applying a series of reflections $R_{i_1}, R_{i_2},\ldots, R_{i_s}$ to (Q,α) , where i_1 is admissible in Q , i_2 is admissible after reversing the arrows with extremity i_1 and so on.

4.2. Real and imaginary roots.

With each vertex i of Q to which no loop is attached we associate a reflection $r_i : \mathbb{Z}^n \to \mathbb{Z}^n$ given by $r_i(\alpha) = \alpha - (\alpha,\epsilon_i)\epsilon_i$. The Weyl group $W = W_Q$ is the subgroup of $GL(\mathbb{Z}^n)$ generated by the r_i . It is contained in the orthogonal group $O(\mathbb{Z}^n, q_Q)$.

A root of Q is a vector $\alpha \in \mathbb{N}^n$ such that $R(Q,\alpha)$ contains an indecomposable representation. Roots have connected support. If for a root α we have $\mu(R(\alpha)_{ind}) \geq 1$, α is called imaginary, and real otherwise. So a root α is real if and only if $R(\alpha)_{ind}$ contains a finite number of orbits. We will see as a consequence of the main theorem that in this case $R(\alpha)_{ind}$ is one single orbit. We denote by $\Delta = \Delta(Q)$, $\Delta_{re} = \Delta(Q)_{re}$ and $\Delta_{im} = \Delta(Q)_{im}$ the sets of all roots, the real roots, and imaginary roots, respectively.

A simple root is a vector ϵ_i, where i is a vertex in which no loop is attached. Equivalently, ϵ_i does not lie in $F = F_Q$ (3.1). The set of simple roots is denoted by $\Pi = \Pi_Q$. Clearly $\Pi \subseteq \Delta_{re}$.

If Q is a Dynkin quiver, then $W\Pi = \Delta \cup -\Delta$ is the corresponding root system and W its Weyl group [B].

4.3. Kac's theorem.

THEOREM. a) $\Delta(Q)_{re} = W\Pi \cap \mathbb{N}^n$; *if* $\alpha \in \Delta(Q)_{re}$, *then* $R(\alpha)_{ind}$ *is one orbit*.

b) $\Delta(Q)_{im} = WF_Q$; *if* $\alpha \in \Delta(Q)_{im}$, *then* $\mu(R(Q,\alpha)_{ind}) = 1 - q(\alpha)$.

The proof rests on the following crucial lemma, which we will prove in chapter 5.

FUNDAMENTAL LEMMA. *For* $\alpha \in \mathbb{N}^n$ *the number of isomorphism classes of indecomposables* V *with* $\dim V = \alpha$ *as well as* $\mu(R(Q,\alpha)_{ind})$ *only depend on the underlying graph* $\bar{Q}$ *of* Q (and not on the orientation).

Proof of the theorem. For $r_i \in W$ and $\alpha \in \Delta\backslash\{\epsilon_i\}$, $r_i\alpha$ lies again in Δ. Indeed, by the fundamental lemma we may assume that i is admissible, apply 4.1 and the fundamental lemma again. So $\Delta\backslash W\Pi$ is stable under W ; we want to show that it is contained in WF, where $F = F_Q$. For $\alpha \in \Delta\backslash W\Pi$ we choose $\beta \in W\alpha$ with minimal height $ht(\beta) = \sum_{i=1}^{n} \beta(i)$. Then

$$ht(r_i\beta) = ht(\beta) - (\beta, \epsilon_i) \geq ht(\beta)$$

for all $r_i \in W$, which implies that $\beta \in F$. Note that $(\beta, \epsilon_j) \leq 0$ for $\epsilon_j \in F$. We conclude that $\Delta \subseteq (W\Pi \cap \mathbb{N}^n) \cup WF$. Conversely, F and thus WF lies in $\Delta\backslash W\Pi$ by theorem 3.3 and remark 3.5.

In order to prove

$$\mu(R(Q,\alpha)_{ind}) = 1 - q(\alpha)$$

for $\alpha \in \Delta(Q)_{im}$, we write $\alpha = r_{i_s} \dots r_{i_1}\beta$, $\beta \in F$, and proceed by induction on s. For $s = 0$, the result follows from theorem 3.4 and remark 3.5. For $s \geq 1$, we set $\gamma = r_{i_{s-1}} \dots r_{i_1}\beta$ and choose an orientation Q' for which i_s is admissible. By Q'' we denote the quiver obtained from Q' by reversing the arrows with extremity i_s. Using the fundamental lemma and 4.1, we find

$$\mu(R(Q,\alpha)_{ind}) = \mu(R(Q',\alpha)_{ind}) = \mu(R(Q'',\gamma)_{ind}) = \mu(R(Q,\gamma)_{ind}),$$

which by the induction hypothesis is equal to

$$1 - q(\gamma) = 1 - q(\alpha).$$

An analogous argument for $\alpha \in W\Pi \cap \mathbb{N}^n$ finishes the proof.

The following proposition from [K1, prop.1.6] generalizes theorems 1 and 2 of chapter 1.

PROPOSITION. Let Q be a connected quiver whose proper subquivers are all either of finite or of tame type. Then

$$\Delta(Q)_{re} = \{\alpha \in \mathbb{N}^n : q(\alpha) = 1\},$$
$$\Delta(Q)_{im} = \{\alpha \in \mathbb{N}^n\setminus\{0\} : q(\alpha) \leq 0\}.$$

Example.

$Q = 1 \rightrightarrows 2$ (two arrows from 1 to 2, one arrow from 2 to 1)

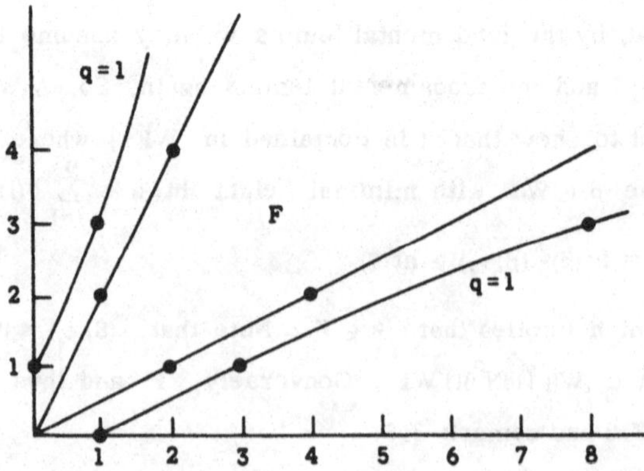

5. PROOF OF THE FUNDAMENTAL LEMMA.

5.1. Starting from a quiver Q, any orientation on $\bar{Q}$ can be obtained in several steps by reversing one arrow at the time. For any arrow $\psi : i \to j$ of Q, we can write

$$R(Q, \alpha) = \bar{R} \times H ,$$

where

$$H = \mathrm{Hom}(k^{\alpha(i)}, k^{\alpha(j)})$$

and

$$\bar{R} = \prod_{\varphi \neq \psi} \mathrm{Hom}(k^{\alpha(t\varphi)}, k^{\alpha(h\varphi)}) .$$

Thus we have to compare the numbers of non-isomorphic indecomposables in $\bar{R} \times H$ and $\bar{R} \times H^*$, where

$$H^* \cong \mathrm{Hom}(k^{\alpha(j)}, k^{\alpha(i)})$$

is the dual vector space of H. Unfortunately, there seems to be no way of doing this directly over an arbitrary algebraically closed field k.

However, a result of Brauer shows that this comparison is possible over finite fields (5.5). Counting points of varieties in finite fields yields the fundamental lemma for the algebraic closure $\overline{\mathbb{F}}_p$ of $\mathbb{F}_p$, for any prime p (5.6). Finally, interpreting the representations in $R(Q, \alpha)$ over an algebraically closed field k as the k-valued points of a scheme over $\mathbb{Z}$, one obtains the lemma in characteristic zero (5.7). So this is one of the examples where the only proof known for a result about fields of characteristic zero passes via fields of positive characteristic.

For any field k -not necessarily algebraically closed- we denote by $R(Q, \alpha)(k)$ the set of k-representations of Q of dimension type α:

$$R(Q, \alpha)(k) = \prod_{\varphi \in Q_1} \mathrm{Hom}_k(k^{\alpha(t\varphi)}, k^{\alpha(h\varphi)}) .$$

Isomorphism classes of k-representations correspond to orbits of $GL(\alpha, k)$ in $R(Q, \alpha)(k)$.

5.2. Fix a prime number p and an algebraic closure k of $\mathbb{F}_p$. Recall that k contains precisely one field $\mathbb{F}_{p^r}$ with p^r elements for $r \in \mathbb{N}$, which is the fixed field in k of the r-th power of the Frobenius automorphism

$x \mapsto x^p$ of k. All fields K, L, E, F occuring up to section 5.6 are finite subfields of k.

Let $K \subset L$ be an extension of degree r, and consider the functors

$$L \otimes_K \ : R(Q,\alpha)(K) \to R(Q,\alpha)(L)$$

and, for a fixed choice of a k-basis $L = K^r$,

$$|K : R(Q,\alpha)(L) \to R(Q, r\alpha)(K) .$$

Obviously $(L \otimes_K V)|K$ is isomorphic to V^r for any representation V in $R(Q,\alpha)(K)$. In combination with the theorem of Krull-Schmidt, this implies the following.

LEMMA 1. <u>If for two representations</u> V_1, V_2 <u>in</u> $R(Q,\alpha)(K)$ <u>the representations</u> $L \otimes_K V_1$ <u>and</u> $L \otimes_K V_2$ <u>are isomorphic in</u> $R(Q,\alpha)(L)$, <u>then</u> V_1 <u>and</u> V_2 <u>are isomorphic in</u> $R(Q,\alpha)(K)$.

Let Γ be the Galois group of L over K, which is cyclic of order r, and denote by $L\Gamma$ the skew group algebra of Γ over L: As a L-(left) vector space, $L\Gamma$ has the elements of Γ as a basis, and

$$x\sigma y\tau = x\,\sigma(y)\,\sigma\tau$$

for x, y in L and σ, τ in Γ. We let Γ operate on $L\Gamma$ by left multiplication. The fixed point set $(L\Gamma)^\Gamma$ for this operation is obviously a K-left and L-right vector space.

LEMMA 2. <u>The map</u>

$$\mu : L \otimes_K (L\Gamma)^\Gamma \to L\Gamma$$

<u>given by the multiplication is an isomorphism of</u> (left and right) L-<u>vector spaces</u>.

<u>Proof</u>. Choosing a normal basis $\{\sigma(x), \sigma \in \Gamma\}$ for L over K, one sees immediately that

$$\dim_K (L\Gamma)^\Gamma = r .$$

Thus the two L-vector spaces $L \otimes_K (L\Gamma)^\Gamma$ and $L\Gamma$ have the same dimension r. The theorem about the linear independence of distinct σ in Γ within $\mathrm{End}_K L$ shows that any L-linear form on $L\Gamma$ which annihilates $\mathrm{im}\,\mu$ is zero. Therefore μ is surjective.

Remark. For any representation W in $R(Q,\alpha)(L)$ we have the decomposition

$$L\Gamma \otimes_L W = \bigoplus_{\sigma \in \Gamma} {}^{\sigma}W$$

of $L\Gamma \otimes_L W$ as a direct sum of L-representations

$${}^{\sigma}W = \{\sigma \otimes w : w \in W\}$$

of dimension type α.

The representation ${}^{\sigma}W$ can be described as follows : for any vertex i of Q, the L-vector space ${}^{\sigma}W(i)$ has the same underlying abelian group as $W(i)$, but the product $a \cdot w$ is given by $\sigma^{-1}(a)w$ for $a \in L$ and $w \in {}^{\sigma}W(i)$; for any arrow φ, ${}^{\sigma}W(\varphi)$ equals $W(\varphi)$. Obviously ${}^{\sigma}W|K$ is isomorphic to $W|K$.

5.3. An indecomposable representation V in $R(Q,\alpha)(K)$ is called absolutely indecomposable if $k \otimes_K V$ is indecomposable. Equivalently, $L \otimes_K V$ is indecomposable for any extension L of K. An extension L of K is called a splitting field for a representation V in $R(Q,\alpha)(K)$ if $L \otimes_K V$ is a direct sum of absolutely indecomposable representations in $R(Q,\alpha)(L)$.

LEMMA 1. *Let V be indecomposable in $R(Q,\alpha)(K)$. Then $L = \operatorname{End} V/\operatorname{rad} \operatorname{End} V$ is a splitting field for V. Moreover $r = [L:K]$ divides α, and we have*

$$L \otimes_K V \simeq L\Gamma \otimes_L W = \bigoplus_{\sigma \in \Gamma} {}^{\sigma}W,$$

where $\Gamma = \operatorname{Gal}(L:K)$, for some absolutely indecomposable representation W in $R(Q,\frac{\alpha}{r})(L)$, and the representations ${}^{\sigma}W$ are pairwise non-isomorphic.

Proof. Since V is indecomposable, $L = \operatorname{End} V/\operatorname{rad} \operatorname{End} V$ is a (finite) skew field and thus a field. We have

$$\operatorname{End}(L \otimes_K V) \simeq L \otimes_K \operatorname{End} V,$$

and also

$$\operatorname{rad} \operatorname{End}(L \otimes_K V) \simeq L \otimes_K \operatorname{rad} \operatorname{End} V,$$

since K is perfect. So we find an isomorphism

$$\operatorname{End}(L \otimes_K V)/\operatorname{rad} \operatorname{End}(L \otimes_K V) \simeq L \otimes_K L \simeq L^r.$$

Therefore $L \otimes_K V$ can be decomposed in $R(Q,\alpha)(L)$ as a direct sum

$$L \otimes_K V = W_1 \oplus \ldots \oplus W_r \,,$$

and the W_i's are pairwise non-isomorphic and have L as endomorphism algebra. As

$$\mathrm{End}(E \otimes W_i)/\mathrm{rad}\ \mathrm{End}(E \otimes W_i) \xrightarrow{\sim} E$$

for any extension field E of L, W_i is absolutely indecomposable for $i = 1, \ldots, r$. For any σ in Γ, the representations $L \otimes_K V$ and

$${}^{\sigma}(L \otimes_K W) = {}^{\sigma}W_1 \oplus \ldots \oplus {}^{\sigma}W_r$$

are isomorphic. Using Krull-Schmidt, we find that $\{W_1, \ldots, W_r\}$ is one Γ-orbit, up to isomorphism. This proves the lemma.

Let W be in $R(Q,\alpha)(L)$. A subfield $K \subseteq L$ is a <u>field of definition</u> for W if there is a representations V in $R(Q,\alpha)(K)$ such that W is induced from V; i.e., W is isomorphic to $L \otimes_K V$.

LEMMA 2. <u>A subfield</u> K <u>of</u> L <u>such that</u> ${}^{\sigma}W$ <u>is isomorphic to</u> W <u>for all</u> $\sigma \in \Gamma = \mathrm{Gal}(L:K)$ <u>is a field of definition for</u> W.

<u>Proof</u>. Clearly we may suppose that the indecomposables in a decomposition of W form one Γ-orbit up to isomorphism. By our hypothesis, $L\Gamma \otimes_L W$ is isomorphic to W^r, and lemma 2 of 5.2 tells us that $L\Gamma \otimes_L W$ is induced from $(L\Gamma)^{\Gamma} \otimes_L W$ in $R(Q, r\alpha)(K)$. Thus we have

$$W^r \xrightarrow{\sim} L \otimes_K V_1 \oplus \ldots \oplus L \otimes_K V_t$$

for some indecomposable K-representations $V_1, \ldots, V_t$. Our assumption on the indecomposables occuring in W implies that $L \otimes_K V_i$ is isomorphic to some W^{s_i} for all i. So we have found an integer s and an indecomposable K-representation V such that W^s is induced from V.

Set $F = \mathrm{End}\, V/\ \mathrm{rad}\ \mathrm{End}\, V$, and let E be a common extension field of F and L in k. By lemma 1,

$$E \otimes_K V \xrightarrow{\sim} E \otimes_L W^s$$

is a direct sum of pairwise non-isomorphic indecomposables, which implies $s = 1$.

COROLLARY. a) Any splitting field E for an indecomposable representation V in $R(Q,\alpha)(K)$ contains $L = \operatorname{End} V/\operatorname{rad} \operatorname{End} V$.

b) Any field of definition K for a representation W in $R(Q,\alpha)(L)$ contains the fixed field L^G, where $G = \{\sigma \in \operatorname{Gal}(L : \mathbb{F}_p) : {}^{\sigma}W \simeq W\}$.

We call the field L of a) the minimal splitting field of V and the field L^G of b) the minimal field of definition of W.

Proof. a) For any splitting field E for V, we have a decomposition

$$E \otimes_K V = U_1 \oplus \ldots \oplus U_r ,$$

where the U_i's are absolutely indecomposable and where $r = [L:K]$, independently of E by lemma 1. As a consequence of lemma 2, the U_i's form one orbit up to isomorphism under $\operatorname{Gal}(E:K)$, since V is indecomposable. Thus r divides $[E:K]$, which implies $L \subseteq E$.

b) If K is a field of definition for W, obviously the group $\operatorname{Gal}(L:K)$ is contained in $G = \operatorname{Gal}(L : L^G)$. Then $[L:K]$ divides $[L : L^G]$, which implies that $[L^G : \mathbb{F}_p]$ divides $[K:\mathbb{F}_p]$, and thus $L^G \subseteq K$.

5.4. We sum up our results in form of the following lemma. For any power q of p we set :

$\nu(Q,\alpha;q)$ = number of isomorphism classes of indecomposable representations in $R(Q,\alpha)(\mathbb{F}_q)$,

$\nu^a(Q,\alpha;q)$ = number of isomorphism classes of absolutely indecomposable representations in $R(Q,\alpha)(\mathbb{F}_q)$,

$\rho(Q,\alpha;q)$ = number of isomorphism classes of absolutely indecomposable representations in $R(Q,\alpha)(\mathbb{F}_q)$ with minimal field of definition $\mathbb{F}_q$.

LEMMA.

a) $\nu^a(Q,\alpha;q) = \sum_{\mathbb{F}_{q'} \subseteq \mathbb{F}_q} \rho(Q,\alpha;q')$

b) $\nu(Q,\alpha;q) = \nu^a(Q,\alpha;q) + \sum_r \frac{1}{r} \rho(Q,\frac{\alpha}{r};q^r)$, where r ranges over all integers greater than 1 for which $\frac{\alpha}{r}$ belongs to $\mathbb{N}^n$.

Proof. a) is clear, and b) says that the isomorphism classes of indecomposables in $R(Q,\alpha)(K)$ with minimal splitting field $\mathbb{F}_{q^r}$ are in bijection with the isomorphism classes of $Gal(\mathbb{F}_{q^r}:\mathbb{F}_q)$ -orbits of indecomposables in $R(Q,\frac{\alpha}{r})(\mathbb{F}_{q^r})$ with minimal field of definition $\mathbb{F}_{q^r}$.

5.5. The "fundamental lemma for finite fields".

Fix a finite field $F \subset k$.

LEMMA. The number of isomorphism classes of representations in $R(Q,\alpha)(F)$ is independent of the orientation of Q.

Proof. Using the notations of 5.1, we have to show that $GL(\alpha,F)$ has the same number of orbits in $\bar{R}\times H$ as in $\bar{R}\times H^*$. Choose representatives $V_1,\dots,V_r$ from the (finitely many !) $GL(\alpha,F)$-orbits in $\bar{R}$. A $GL(\alpha,F)$-orbit in $\bar{R}\times H$ or $\bar{R}\times H^*$ corresponds to the choice of a V_i and of a $C_{GL(\alpha,F)}V_i$-orbit in H or H^*, respectively.

Choose i in $\{1,\dots,r\}$, and set $G = C_{GL(\alpha,F)}V_i$. Consider the $\mathbb{C}$-vector space $\mathbb{C}^H$ of maps from H to $\mathbb{C}$, and let G operate on $\mathbb{C}^H$ by

$$(g\cdot f)(h) = f(g^{-1}\cdot h) .$$

A fixed point of G in $\mathbb{C}^H$ is a map which is constant on G-orbits in H. Thus the number of G-orbits in H equals the dimension of the vector space $(\mathbb{C}^H)^G$ of fixed points of G in $\mathbb{C}^H$.

To prove our assertion, it suffices to exhibit a G-equivariant isomorphism

$$F : \mathbb{C}^H \to \mathbb{C}^{H^*} .$$

The Fourier transform, which is defined as follows, does the trick :

$$(Ff)(\varphi) = \frac{1}{|H|} \sum_{h\in H} f(h) \exp\left(\frac{2\pi i}{p} tr(\varphi(h))\right)$$

for $f \in \mathbb{C}^H$, $\varphi \in \mathbb{C}^{H^*}$. Here $|H|$ is the cardinality of H, and $tr : F \to \mathbb{F}_p$ the usual trace map. An easy computation shows that

$$(F(Ff))(h) = \frac{1}{|H|} f(-h) .$$

So F is an isomorphism, which is G-equivariant by definition.

Remark. The result that a finite group operating linearly on a finite vector space V has the same number of orbits in V and V^* seems to be due to Brauer, but we have not been able to find a precise reference.

COROLLARY 1. The number of isomorphism classes of indecomposable representations in $R(Q,\alpha)(F)$ is independent of the orientation of Q.

Proof. Use induction on $ht(\alpha) = \sum_{i=1}^{n} \alpha(i)$ and Krull-Schmidt.

COROLLARY 2. The numbers $\nu^a(Q,\alpha;q)$ and $\rho(Q,\alpha;q)$ defined in 5.4 are independent of the orientation of Q for any $q = p^s$ and any α.

Proof. This is clearly true for any q if $ht(\alpha) = 1$. Suppose it is true for any q and any β with $ht(\beta) < ht(\alpha)$. Then formula b) of Lemma 5.4 implies that $\nu^a(Q,\alpha;q)$ is independent of the orientation, as $\nu(Q,\alpha;q)$ is by corollary 1.

To prove the independence of $\rho(Q,\alpha;p^s)$ of the orientation, we keep α fixed and apply induction on s. For $s = 1$, we have

$$\rho(Q,\alpha;p) = \nu^a(Q,\alpha;p) .$$

The general case follows from formula a) of lemma 5.4, which says that

$$\nu^a(Q,\alpha;p^s) = \rho(Q,\alpha;p^s) + \sum_{\substack{t/s \\ t\neq s}} \rho(Q,\alpha;p^t) .$$

5.6. The fundamental lemma for $k = \overline{\mathbb{F}}_p$.

Let $\mathcal{U} \subseteq k^N$ be a locally closed subset which is stable under the Frobenius automorphism σ of k^N. For any finite subfield $\mathbb{F}_q$ of k, set

$$\mathcal{U}(\mathbb{F}_q) = \mathcal{U} \cap \mathbb{F}_q^N .$$

This set may well be empty for small q's. However, the famous result of Lang and Weil about counting points of varieties in finite fields ([LW], see also [Sch] for an elementary approach) implies the cardinality of $\mathcal{U}(\mathbb{F}_q)$

behaves as follows :

PROPOSITION. If $\mathcal{U}$ is stable under σ , then

$$\# \mathcal{U}(\mathbb{F}_q) \approx c\, q^{\dim \mathcal{U}}$$

where c is the number of irreducible components of $\mathcal{U}$ of maximal dimension.

The symbol $f(q) \approx g(q)$ means that either $f(q) = g(q) = 0$ for q large or that $f(q), g(q) \neq 0$ for q large and $\frac{f(q) - g(q)}{f(q)} \xrightarrow[q\to\infty]{} 0$.

The set $R(Q,\alpha)^{(d)}_{\mathrm{ind}}$ of indecomposable representations of Q over k of dimension type α and with d-dimensional endomorphism algebra is a locally closed subset of k^N , where $N = \sum_{\varphi \in Q_1} \alpha(t\varphi)\alpha(h\varphi)$ (2.5) . Let V be in $R(Q,\alpha)^{(d)}_{\mathrm{ind}}$. All coefficients of all matrices $V(\varphi)$, $\varphi \in Q_1$, lie in some finite subfield $K \subset k$, and we may view V as an absolutely indecomposable representation in $R(Q,\alpha)(K)$ with d-dimensional endomorphism algebra over K . An easy computation shows that applying the Frobenius σ to all coefficients of all matrices $V(\varphi)$ while keeping the bases of all $V(i)$ fixed yields a representation isomorphic to ${}^{\sigma}V$ (5.2, remark), where σ is viewed in $\mathrm{Gal}(K : \mathbb{F}_p)$. Therefore $R(Q,\alpha)^{(d)}_{\mathrm{ind}}$ is stable under σ , and we may apply the preceding proposition. We find :

$$\# R(Q,\alpha)^{(d)}_{\mathrm{ind}}(\mathbb{F}_q) \approx c(d)\, q^{\mu(d) + \dim GL(\alpha) - d} ,$$

where $\mu(d)$ denotes the number of parameters (3.4) and $c(d)$ the number of irreducible components of maximal dimension of $R(Q,\alpha)^{(d)}_{\mathrm{ind}}$.

For an absolutely indecomposable representation V in $R(Q,\alpha)(\mathbb{F}_q)$, we have

$$GL(\alpha)\cdot V \cap \mathbb{F}_q^N = GL(\alpha, \mathbb{F}_q)\cdot V$$

by lemma 1 of 5.2. If V lies in $R(Q,\alpha)^{(d)}_{\mathrm{ind}}$, its $GL(\alpha)$-orbit is a locally closed irreducible σ-stable subset of dimension $GL(\alpha) - d$, and we find :

$$\# (GL(\alpha)V)(\mathbb{F}_q) \approx q^{\dim GL(\alpha) - d} .$$

As a consequence, the number of $GL(\alpha, \mathbb{F}_q)$-orbits in $R(Q,\alpha)^{(d)}_{\mathrm{ind}}(\mathbb{F}_q)$

behaves like $c(d)\,q^{\mu(d)}$, and thus the number $\nu^a(Q,\alpha;q)$ of $GL(\alpha,\mathbb{F}_q)$-orbits in $R(Q,\alpha)_{ind}(\mathbb{F}_q)$ like cq^{μ}, where $\mu = \max_d \mu(d)$ is the number of parameters of $R(Q,\alpha)_{ind}$ (3.4) and $c = \sum_{\mu(d)=\mu} c(d)$.

But $\nu^a(Q,\alpha;q)$ is independent of the orientation of Q (5.5). Therefore the number of parameters μ and the number c of irreducible components of $R(Q,\alpha)_{ind}$ with number of parameters μ are independent as well. Since for $\mu = 0$, c is the number of orbits, this ends the proof.

Remark. As we may now apply Kac's theorem to $\overline{\mathbb{F}}_p$, we see that $\mu = 1-q(\alpha)$, independently of p. Moreover, there exists precisely one irreducible component of $R(Q,\alpha)_{ind}$ for which the number of parameters is maximal, which means that $c = 1$. Indeed, this is true for $\alpha \in F_Q$, it remains true if we replace (Q,α) by (Q^*,α^*) for some reflection at an admissible vertex (4.1), and it still holds when we change orientation. In particular, if the generic representation of dimension type α is indecomposable, the number $1-q(\alpha)$ of parameters of $R(\alpha)^{max}$ (corollary 2.7) is strictly greater than the one of any other irreducible component of $R(\alpha)_{ind}$.

5.7. End of the proof.

In order to transfer our results in characteristic p to characteristic zero, we have to define our varieties over $\mathbb{Z}$. Let $\mathbb{Z}[X_{\varphi;st}]$ be the polynomial ring over $\mathbb{Z}$ in the variables $X_{\varphi;st}$, where φ ranges over the arrows of Q and $1 \le s \le \alpha(t\varphi)$, $1 \le t \le \alpha(h\varphi)$, and set

$$\mathcal{R} = \mathcal{R}(Q,\alpha) = \operatorname{Spec} \mathbb{Z}[X_{\varphi;st}] .$$

Obviously $\mathcal{R}$ is affine N-space over $\mathbb{Z}$ with $N = \sum_{\varphi\in Q_1} \alpha(h\varphi)\,\alpha(t\varphi)$. For any (commutative !) ring A, the A-valued points $\mathcal{R}(A)$ of $\mathcal{R}$ can be identified with

$$\prod_{\varphi\in Q_1} \operatorname{Hom}_A(A^{\alpha(t\varphi)}, A^{\alpha(h\varphi)}) .$$

In particular, if k is an algebraically closed field, we have

$$\mathcal{R}(k) = R(Q,\alpha)(k) .$$

Next we want to define a scheme whose k-valued points are pairs (V, f) consisting of a representation V and an endomorphism f of V. To this end, we set

$$\mathcal{M} = \operatorname{Spec} \mathbb{Z}[Y_{i;jk}] ,$$

where i ranges over the points of Q and $1 \le j, k \le \alpha(i)$. We denote by Y_i the $\alpha(i) \times \alpha(i)$-matrix whose entries are the $Y_{i;jk}$ and by X_φ the $\alpha(h\varphi) \times \alpha(t\varphi)$ - matrix with entries $X_{\varphi;st}$. For each $\varphi \in Q_1$, the equation

$$X_\varphi \cdot Y_{t\varphi} - Y_{h\varphi} \cdot X_\varphi = 0$$

yields $\alpha(t\varphi)\alpha(h\varphi)$ linear equations in

$$\mathbb{Z}[X_{\varphi;st}] \otimes_{\mathbb{Z}} \mathbb{Z}[Y_{i;jk}] = \mathbb{Z}[X_{\varphi;st}, Y_{i;jk}] .$$

Denote by $\mathcal{J}$ the ideal generated by these linear equations, and let

$$\mathcal{Z} = \operatorname{Spec} k[X_{\varphi;st}, Y_{i;jk}] / \mathcal{J}$$

be the corresponding closed subscheme of $\mathcal{R} \times \mathcal{M}$. Obviously we have for any algebraically closed field

$$\mathcal{Z}(k) = \{(V, f) : V \in R(Q, \alpha)(k), f \in \operatorname{End} V\} .$$

Consider the first projection

$$\pi : \mathcal{Z} \to \mathcal{R}$$

and define the subset

$$\mathcal{R}^{(d)} = \{x \in \mathcal{R} : \dim \pi^{-1}(x) = d\} .$$

Since sending x to (x, 0) is a section for π and since all fibers of π are irreducible, being affine spaces, $\mathcal{R}^{(d)}$ is locally closed by Chevalley's theorem [EGA IV, 13.1]. We endow $\mathcal{R}^{(d)}$ with the structure of a reduced scheme. Clearly we have

$$\mathcal{R}^{(d)}(k) = R^{(d)}(Q, \alpha)(k)$$

for any algebraically closed field k.

In order to define the "subscheme of indecomposables in $\mathcal{R}^{(d)}$", we use the characterization of 2.5 . Let

$$\mathcal{J} \subseteq \mathbb{Z}[Y_{i;jk}]$$

be the ideal generated by the equations which express that Y_i is nilpotent, or equivalently that $Y_i^{\alpha(i)} = 0$, for all vertices i . Then

$$\mathcal{N} = \operatorname{Spec} \mathbb{Z}[X_{\varphi;st}, Y_{i;jk}] / \mathcal{I}+\mathcal{J}$$

is a closed subscheme of $\mathcal{Z}$. Applying, Chevalley's theorem to the first projection

$$\pi' : \mathcal{N} \to \mathcal{R} ,$$

we see that the subset

$$\mathcal{S}^{(t)} = \{ x \in \mathcal{R} : \dim \pi'^{-1}(x) \geq t \}$$

of $\mathcal{R}$ is closed for all t . Indeed, the zero section for π' meets all irreducible components of each fiber. We give the intersection

$$\mathcal{R}_{ind}^{(d)} = \mathcal{R}^{(d)} \cap \mathcal{S}^{(d-1)} ,$$

which is closed in $\mathcal{R}^{(d)}$, the structure of a reduced scheme. For any algebraically closed field k , $\mathcal{R}_{ind}^{(d)}(k)$ consists of those representations with d-dimensional endomorphism algebra for which the nilpotent endomorphism form a subvariety of dimension $d-1$; this is just $R(Q,\alpha)_{ind}^{(d)}(k)$.

Summing up, we have defined for each d a locally closed reduced subscheme $\mathcal{R}^{(d)}$ of $\mathcal{R}$ and a closed reduced subscheme $\mathcal{R}_{ind}^{(d)}$ of $\mathcal{R}^{(d)}$ such that

$$\mathcal{R}^{(d)}(k) = R(Q,\alpha)^{(d)}(k) \quad \text{and} \quad \mathcal{R}_{ind}^{(d)}(k) = R(Q,\alpha)_{ind}^{(d)}(k)$$

for any algebraically closed field k .

Let k be algebraically closed and choose an algebraic closure k_0 within k of the prime subfield of k . For any d , the varieties $\mathcal{R}_{ind}^{(d)}(k)$ and $\mathcal{R}_{ind}^{(d)}(k_0)$ have the same dimension and the same number of irreducible components of maximal dimension [EGA IV, 4.4] . Therefore the fundamental lemma is true for k if it holds for k_0 . It remains to prove it for an algebraic closure of $\mathbb{Q}$.

Consider the canonical morphism

$$f_d : \mathcal{R}_{ind}^{(d)} \to \operatorname{Spec} \mathbb{Z} .$$

There is a non-empty open subset U_d of $\mathbb{Z}$ such that

$$\dim f_d^{-1}(0) = \dim f_d^{-1}(p)$$

for all p in U_d ([EGA IV, 9.2]. If $\bar{\mathbb{Q}}$ and $\bar{\mathbb{F}}_p$ denote algebraic closures of $\mathbb{Q}$ and $\mathbb{F}_p$, then $\mathcal{R}_{ind}^{(d)}(\bar{\mathbb{Q}})$ and $\mathcal{R}_{ind}^{(d)}(\bar{\mathbb{F}}_p)$ are the varieties of $\bar{\mathbb{Q}}$-valued points of $f_d^{-1}(0)$ and $\bar{\mathbb{F}}_p$-valued points of $f_d^{-1}(p)$, respectively. We find

$$\dim \mathcal{R}_{ind}^{(d)}(\bar{\mathbb{Q}}) = \dim \mathcal{R}_{ind}^{(d)}(\bar{\mathbb{F}}_p)$$

for all p in the open set $U = \bigcap_d U_d$ and all d.

For any algebraically closed field k, the number of parameters of $R(Q,\alpha)_{ind}(k)$ is defined as

$$\mu(k) = \max_d (\dim R(Q,\alpha)_{ind}^{(d)}(k) - \dim GL(\alpha) + d) \qquad \text{(see 3.4)}.$$

As $\mu(\bar{\mathbb{F}}_p) = 1 - q(\alpha)$ for all p, we conclude that $\mu(\bar{\mathbb{Q}}) = 1 - q(\alpha)$, independently of the orientation of Q.

Suppose $1 - q(\alpha) = 0$. For any prime p there is a unique d_p such that $R(Q,\alpha)_{ind}^{(d_p)}(\bar{\mathbb{F}}_p)$ is non-empty, and then it is a single $GL(\alpha, \bar{\mathbb{F}}_p)$-orbit. As $d_p = d$ is constant on U, we see that $R(Q,\alpha)_{ind}^{(d')}(\bar{\mathbb{Q}})$ is empty for $d' \neq d$. Moreover, $\mathcal{R}_{ind}^{(d)}(\bar{\mathbb{Q}})$ is connected by [EGA IV, 9.7], and thus consists of a single orbit. This ends the proof of the fundamental lemma for $\bar{\mathbb{Q}}$.

<u>Remark</u>. Refining the last argument one can show that for any algebraically closed field k the number $1 - q(\alpha)$ of parameters is reached for precisely one irreducible component of precisely one $R_{ind}^{(d)}(Q,\alpha)(k)$.

REFERENCES.

[BGP] I. N. BERNSTEIN, I. M. GELFAND and V. A. PONOMAREV. Coxeter functors and Gabriel's theorem. Uspechi Mat. Nauk. 28 (1973), 19-33 ; translated in Russian Math. Surveys 28 (1973), 17-32.

[B] N. BOURBAKI. Groupes et algèbres de Lie, Hermann. Paris.

[DR] V. DLAB and C. M. RINGEL. Indecomposable representations of graphs and algebras. Memoirs of the AMS, vol. 6, Nr. 173 (1976).

[DF] P. DONOVAN and M. R. FREISLICH. The representation theory of finite graphs and associated algebras. Carleton Math. Lecture Notes Nr. 5 (1973).

[G1] P. GABRIEL. Unzerlegbare Darstellungen I. Manuscripta Math. 6 (1972), 71-103.

[G2] P. GABRIEL. Représentations indécomposables. Séminaire Bourbaki, Exposé 444 (1974), in Springer Lecture Notes 431 (1975), 143-169.

[EGA] A. GROTHENDIECK. Eléments de géométrie algébrique IV, Etude locale des schémas et des morphismes de schémas. IHES, Publications mathématiques Nr. 24 (1965) and Nr. 28 (1966).

[K1] V. KAC. Infinite root systems, representations of graphs and invariant theory. Invent. Math. 56 (1980), 57-92.

[K2] V. KAC. Infinite root systems, representations of graphs and invariant theory. J. of Algebra 78 (1982), 141-162.

[K3] V. KAC. Root systems, representations of quivers and invariant theory, in Proceedings of the CIME, Montecatini 1982. Springer Lecture Notes 996, 74-108.

[K4] V. KAC. Inifinite dimensional Lie algebras, an introduction. Progress in Mathematics 44 (1983), Birkhäuser.

[Kr1] H. KRAFT. Parametrisierung von Konjugationsklassen in $\mathfrak{sl}_n$. Math. Ann. 234 (1978), 209-220.

[Kr2] H. KRAFT. Geometrische Methoden in der Invariantentheorie. Aspekte der Mathematik (1984), Vieweg.

[LW] S. LANG and A. WEIL. Numbers of points of varieties in finite fields. Am. J. of Math. 76 (1954), 819-827.

[L] G. LUSZTIG. On the finiteness of the number of unipotent classes. Invent. Math. 34 (1976), 201-213.

[N] L. A. NAZAROVA. Representations of quivers of infinite type. Math. USSR, Izvestija, Ser. Mat. 37 (1973), 752-791.

[P] D. PETERSON. Geometry of the adjoint representation of a complex semisimple Lie algebra. Thesis, Harvard University 1978.

[SK] M. SATO and T. KIMURA. A class of irreducible prehomogeneous vector spaces and their relative invariants. Nagoya Math. J. 65 (1977), 1-155.

[Sch] W. SCHMIDT. Equations over finite fields, an elementary approach. Springer Lecture Notes 536 (1976).

[V] E. B. VINBERG. Discrete linear groups generated by reflections. Math. USSR, Izvestija, Ser. Mat. 5 (1971), 1083-1119.

- o -

REFERENCES

[BGP] I.N. BERNSTEIN, I.M. GELFAND and V.A. PONOMAREV, Coxeter functors and Gabriel's theorem, Uspehi Mat. Nauk. 28 (1973), 19-33 (= English translation: Russian Math. Surveys 28 (1973), 17-32).

[B] N. BOURBAKI, Groupes et algèbres de Lie, Hermann, Paris.

[DR] V. DLAB and C.M. RINGEL, Indecomposable representations of graphs and algebras, Memoirs of the AMS, vol. 6, No. 173 (1976).

[DG] P. DONOVAN and M.R. FREISLICH, The representation theory of finite graphs and associated algebras, Carleton Math. Lecture Notes No. 5 (1973).

[G1] P. GABRIEL, Unzerlegbare Darstellungen I, Manuscripta Math. 6 (1972), 71-103.

[G2] P. GABRIEL, Représentations indécomposables, Séminaire Bourbaki, Exposé 444 (1974), in Springer Lecture Notes 431 (1975), 143-169.

[EGA] A. GROTHENDIECK, Eléments de géométrie algébrique IV, Étude locale des schémas et des morphismes de schémas, IHES, Publications mathématiques Nr. 24 (1965) and Nr. 28 (1966).

[K1] V. KAC, Infinite root systems, representations of graphs and invariant theory, Invent. Math. 56 (1980), 57-92.

[K2] V. KAC, Infinite root systems, representations of graphs and invariant theory II, J. of algebra 78 (1982), 141-162.

[K3] V. KAC, Root systems, representations of quivers and invariant theory, in Proceedings of the CIME, Montecatini 1982, Springer Lecture Notes 996, 74-108.

[L-4] [illegible] LAMOTKE, Infinite dimensional Lie algebras, An introduction, Progress in Mathematics 44 (1983), Birkhäuser.

[Kr] H. KRAFT, Parametrisierung von Konjugationsklassen in $\mathfrak{sl}_n$, Math. Ann. 234 (1978), 209-220.

[K-3] H. KRAFT, Geometrische Methoden in der Invariantentheorie, Aspekte der Mathematik D1, Vieweg.

[LW] S. LANG and A. WEIL, Numbers of points of varieties in finite fields, Am. J. of Math. 76 (1954), 819-827.

[L] [illegible] LUSZTIG, On the finiteness of the number of unipotent classes, Invent. Math. 34 (1976), 201-213.

[N] L. A. NAZAROVA, Representations of quivers of infinite type, Izv. Akad. Nauk SSSR, Ser. Mat. 37 (1973), 752-791.

[P] D. PETERSON, Geometry of the adjoint representation of a complex semisimple Lie algebra, Thesis, Harvard University 1978.

[SK] M. SATO and T. KIMURA, A classification of irreducible prehomogeneous vector spaces and their relative invariants, Nagoya Math. J. 65 (1977), 1-155.

[Sl] P. SLODOWY, Simple singularities and simple algebraic groups, Springer Lecture Notes 815 (1980).

[V] E. B. VINBERG, Discrete linear groups generated by reflections, Math. USSR Izvestija, Ser. Mat. 5 (1971), 1083-1119.

COHEN-MACAULAY MODULES ON HYPERSURFACE SINGULARITIES

H. Knörrer
Mathematisches Institut der Universität Bonn, Wegelerstr. 10,
5300 Bonn 1, Germany

In the last few years the maximal Cohen-Macaulay modules over local rings of singularities have been studied by methods of representation theory, commutative algebra, and algebraic geometry. The article of M. Auslander in this volume and the present article are intended as a survey over some of the recent progress in this subject. This part of the survey will be concerned almost exclusively with the situation of complex hypersurface singularities. In fact it is centered around the following result from [Buchweitz-Greuel-Schreyer], [Knörrer]:

Theorem: Let R be the local ring of a (complex) hypersurface singularity. There are finitely many isomorphism classes of indecomposable maximal Cohen-Macaulay modules over R if and only if R is the local ring of a simple singularity.

In the first part I want to give a brief account of the role the simple singularities play in the classification of hypersurface singularities. Of course this can be only a very rough and subjective sketch; a much broader and more competent description can be found e.g. in [Arnold], [Arnold et al.], [Durfee], [Slodowy]. The second chapter mainly reports on the group-theoretic and algebro-geometric description of maximal Cohen-Macaulay modules over two-dimensional simple hypersurface singularities. Chapter 3 contains a brief sketch of the proof of the theorem stated above.

In preparing these notes I profited very much from expositions of this and related material which R. Buchweitz has given at conferences in Göttingen and Oberwolfach.

1. SIMPLE HYPERSURFACE SINGULARITIES

1.1 Notation

The theory of hypersurface singularities is concerned with the "local" behaviour of holomorphic functions $f : U \to \mathbb{C}$ on some open subset U of $\mathbb{C}^n$ resp. the structure of their zero-sets

$$V(f) := \{x \in U | f(x) = 0\}$$

Sets of the form $V(f)$ as above are usually called <u>(analytic) hypersurfaces in U.</u> To give the term "local" above a precise meaning one introduces the concept of a germ of a hypersurface:

<u>Definition:</u> A <u>germ</u> of an analytic hypersurface in $p \in \mathbb{C}^n$ is an equivalence class of pairs (U,V), where U is an open subset of $\mathbb{C}^n$ containing p, V an analytic hypersurface in U which passes through p, subject to the following equivalence relation: (U_1,V_1) and (U_2,V_2) are equivalent if and only if there is a neighbourhood U of p in $\mathbb{C}^n$ which is contained in $U_1 \cap U_2$ and such that $U \cap V_1 = U \cap V_2$. We write (V,p) for such a germ of a hypersurface in p. In studying such germs of analytic hypersurfaces we may assume that p is the origin $0 \in \mathbb{C}^n$. Then one easily sees that there is a natural bijection between

a) germs of hypersurfaces in $0 \in \mathbb{C}^n$

b) reduced principal ideals in the ring $\mathbb{C}\{x_1,\ldots,x_n\}$ of convergent power series.

(To go from (b) to (a) one chooses a generator f of the ideal. f defines a hypersurface in some neighbourhood of 0, and we associate to the ideal the germ $(V(f),0)$. The fact that this construction yields a bijection between (a) and (b) follows from an analytic version of Hilbert's Nullstellensatz (cf. [Gunning-Rossi] II.E).

There is an obvious geometric notion of isomorphisms of germs; and it turns out that up to isomorphism a germ is uniquely determined by its local ring

$$\mathcal{O}_{(V,0)} := \mathbb{C}\{x_1,\ldots,x_n\}/I$$

where I is the principal ideal associated to $(V,0)$. If this local ring is regular (i.e. if the differential of a generator of I does not vanish at 0) we say that $(V,0)$ is smooth (or regular). A germ of a hypersurface that is not smooth is called a hypersurface singularity.

1.2 Families of Hypersurface Singularities:

In many geometric situations - e.g. in the so-called catastrophy theory or in moduli questions in algebraic geometry - one considers not only one individual hypersurface, but families of hypersurfaces that are parametrised by some analytic variety. The way a germ of a hypersurface can fit (locally) in such a family is described by the concept of "unfolding" and "deformation".

Definition: Let $f : U \to \mathbb{C}$ be a holomorphic function on a neighbourhood U of $0 \in \mathbb{C}^n$ with $f(0) = 0$. An unfolding of f is a holomorphic map $F : U \times T \to \mathbb{C}$, where T is a neighbourhood of 0 in some $\mathbb{C}^k$,

$$F(x,0) = f(x) \qquad \text{for all } x \in U$$
$$F(0,t) = 0 \qquad \text{for all } t \in T.$$

An unfolding thus gives a family $(f_t)_{t\in T}$ of functions with $f_0 = f$ (namely $f_t(x) := F(x,t)$). The associated family of germs of hypersurfaces $(V(f_t),0)_{t\in T}$ is called a deformation (with section) of $(V(f),0)$*.

Examples:

1.) $$f(x_1,x_2) = x_1^3 + x_2^2, \qquad F(x_1,x_2,t) = x_1^3 - tx_1^2 + x_2^2$$

One can show that for $t \neq 0$ the germ $(V(f_t),0)$ is an ordinary double point, i.e. $\mathcal{O}_{(V(f_t),0)} \cong \mathbb{C}\{z_1,z_2\}/(z_1^2 + z_2^2)$ for all $t \neq 0$.

2.) $$f(x_1,x_2) = x_1x_2(x_1^2 - x_2^2)$$

$$f_t(x_1,x_2) = x_1x_2(x_1 + x_2)(x_1 - (t+1)x_2).$$

---

*This definition is ad hoc and too simple minded from a functorial point of view, for a more systematic treatment see e.g. [Looijenga] 6.C

In this deformation one sees that for almost all $(t_1,t_2) \in \mathbb{C}\times\mathbb{C}$ the germs $(V(f_{t_1}),0)$, $(V(f_{t_2}),0)$ are not isomorphic. More precisely one has $(V(f_{t_1}),0) \cong (V(f_{t_2}),0)$ if and only if the sets $\{0,-1,t_1+1,\infty\}$ and $\{0,-1,t_2+1,\infty\}$ have equivalent cross-ratios.

For isolated hypersurface singularities there is an effective procedure to "classify" all possible deformations, see e.g. [Arnold et al.], [Damon], [Looijenga], [Wall]. (A hypersurface singularity in $0 \in \mathbb{C}^n$ is called isolated if it has a representative (U,V) such that (V,x) is smooth for all $x \in V - \{0\}$. This is equivalent to demanding that the localization of $\mathcal{O}_{(V,0)}$ at any non-maximal prime ideal is regular.)

1.3 The Classification of Simple Singularities:

There are various principles according to which singularities can be classified. One of the most fruitful approaches has been developed around 1968 by [Arnold]; it is based on the "deformation hierarchy" of singularities.

Definition (Arnold): A hypersurface singularity $(V,0)$ is called simple (or of modularity zero) if there exists a finite list $(\tilde{V}_1,0),\ldots(\tilde{V}_r,0)$ of germs of hypersurfaces such that for any deformation $(V_t,0)_{t\in T}$ of $(V,0)$ there is a neighbourhood $T' \subset T$ of 0 in T such that for each $t \in T'$ the germ $(V_t,0)$ is isomorphic to one of the germs $(\tilde{V}_i,0)$.

So the singularity of example 2 in (1.2) is not simple. In a similiar way as above one can define singularities of higher modularity, and [Arnold] classified the singularities up to modularity two. Of particular interest to us is the list of simple hypersurface singularities:

Theorem ([Arnold]) Up to isomorphism the simple hypersurface singularities in $0 \in \mathbb{C}^n$ are the ones given by the following equations

$$x_1^{k+1} + x_2^2 + x_3^2 + \ldots + x_n^2 = 0 \qquad (A_k)$$

$$x_1^2 x_2 + x_2^{k-1} + x_3^2 + \ldots + x_n^2 = 0 \qquad (D_k)$$

$$x_1^3 + x_2^4 + x_3^2 + \ldots + x_n^2 = 0 \qquad (E_6)$$

$$x_1^3 + x_1 x_2^3 + x_3^2 + \ldots + x_n^2 = 0 \qquad (E_7)$$

$$x_1^3 + x_2^5 + x_3^2 + \ldots + x_n^2 = 0 \qquad (E_8)$$

The labelling $(A_k, D_k, E_6, E_7, E_8)$ will be explained in ch.2. Observe that all these equations are of the form $f(x_1, x_2) + x_3^2 + \ldots + x_n^2 = 0$, where $f(x_1, x_2) = 0$ defines a simple plane curve singularity. The simple hypersurface singularities have many different geometric characterisations (see e.g. [Durfee]), one of the most beautiful is the following:

Remark: Let G be a finite subgroup of $SL(2, \mathbb{C})$. Its action on $\mathbb{C}^2$ induces an action on the power series ring $\mathbb{C}\{x,y\}$. Then the invariant ring $\mathbb{C}\{x,y\}^G$ is isomorphic to the local ring of a two-dimensional hypersurface singularity. Every two-dimensional hypersurface singularity arises in this way.

The equation of a two-dimensional simple hypersurface thus is the relation between three generators of the invariant ring $\mathbb{C}\{x,y\}^G$. These relations had been studied by H.A. Schwarz and F. Klein in the last century, and consequently the two-dimensional simple hypersurface singularities are sometimes called Kleinian singularities (see also [Slodowy]).

2. HYPERSURFACE SINGULARITIES OF DIMENSION ONE AND TWO

2.1 Maximal Cohen-Macaulay Modules:

There is probably no chance to classify all modules over a positive - dimensional local ring; and we will deal only with a very restricted and geometrically particularly interesting class of modules.

Definition: Let $R = \mathcal{O}_{(V,O)}$ be the local ring of a germ of a hypersurface in $O \in \mathbb{C}^n$. An R-module M is called a maximal Cohen-Macaulay module (MCM) if for some (and then for any) regular subring $P \subset R$ with $[R:P] < \infty$ the module M is free over P.

In particular MCM's over the local ring of a smooth germ are always free. In dimension one and two MCM's are of considerable geometric significance: If $\dim R = 1$ then an R-module is MCM if and only if it is torsion free. Torsion-free modules over curve singularities play an important role in the compactification of moduli spaces of vector bundles over algebraic curves [Newstead] 5,§7. If $\dim R = 2$ then an R-module M is MCM if and only if it is reflexive, i.e. if $\mathrm{Hom}_R(\mathrm{Hom}_R(M,R),R) \cong M$. In the case that (V,O) is an isolated surface singularity then MCM's over R are in a one-to-one correspondence with vectorbundles on $V' - \{O\}$, where V' is a suitable Stein-representative of (V,O).

2.2 Group Representations:

We mentioned in (1.3) that the local rings of the two-dimensional simple hypersurface singularities are all of the form $\mathbb{C}\{x,y\}^G$ where G is a finite subgroup of $SL(2,\mathbb{C})$. This makes it possible to describe all MCM's on these singularities.

Theorem [Herzog]: Let G be a finite subgroup of $GL(2,\mathbb{C})$, $R := \mathbb{C}\{x,y\}^G$. If $\rho : G \to GL(E)$ is a complex representation of G then the "invariant module" $M_\rho := (\mathbb{C}\{x,y\} \otimes E)^G$ is MCM over R. Conversely each MCM over R is isomorphic to one of the modules M_ρ.

This shows that the invariant rings $\mathbb{C}\{x,y\}^G$, $G \subset GL(2,\mathbb{C})$ finite, admit only finitely many isomorphism classes of indecomposable MCM's (we say they are of finite CM-type), namely those which correspond to indecomposable representations of G. The converse to this result is

Theorem (Artin-Verdier, [Auslander], [Esnault]: Let R be the local ring of a normal surface singularity which is of finite CM-type. Then it is a quotient singularity, i.e. there is a finite subgroup $G \subset GL(2,\mathbb{C})$ s.t. $R \cong \mathbb{C}\{x,y\}^G$.

It is well known that the only hypersurface singularities among the two-dimensional quotient singularities are just the simple singularities (cf. [Durfee]). The Auslander-Reiten quivers of the category of MCM's over these singularities has been computed by [M. Auslander], see also [Auslander-Reiten]:

Theorem: Let G be a finite subgroup of $SL(2,\mathbb{C})$, $R := \mathbb{C}\{x,y\}^G$, and let c be the representation of G given by the embedding of G in $SL(2,\mathbb{C})$. If ρ,ρ' are indecomposable representations of G then there exists an irreducible morphism $M_\rho \to M_{\rho'}$ if and only if ρ' is a direct summand of $\rho \otimes c$.

This tensor product with c had been studied by [McKay] and later [Steinberg], and from their description follows.

Corollary: The Auslander-Reiten quiver of a simple two-dimensional hypersurface singularity is an extended Dynkin quiver of type A,D,E. (The individual types are marked in the list of simple singularities in (1.3)).

2.3 Vectorbundles on the Desingularisation:

The fact that the Kleinian singularities were related to Dynkin-diagrams of type A,D,E was - in principle - known since P. Du Val in the 1930's had computed their minimal desingularisation (see [Slodowy]):

Definition: Let V be an analytic hypersurface (more generally any analytic variety) in an open subset of $\mathbb{C}^n$ with 0 as only singular point. A desingularisation (or resolution) of V is a proper holomorphic map $\pi: \tilde{V} \to V$ where

(i) $\tilde{V}$ is a complex manifold (i.e. smooth!)

(ii) π induces an isomorphism between $\tilde{V} - \pi^{-1}(0)$ and $V - \{0\}$

(iii) $\pi^{-1}(0)$ is a closed subvariety with empty interior.

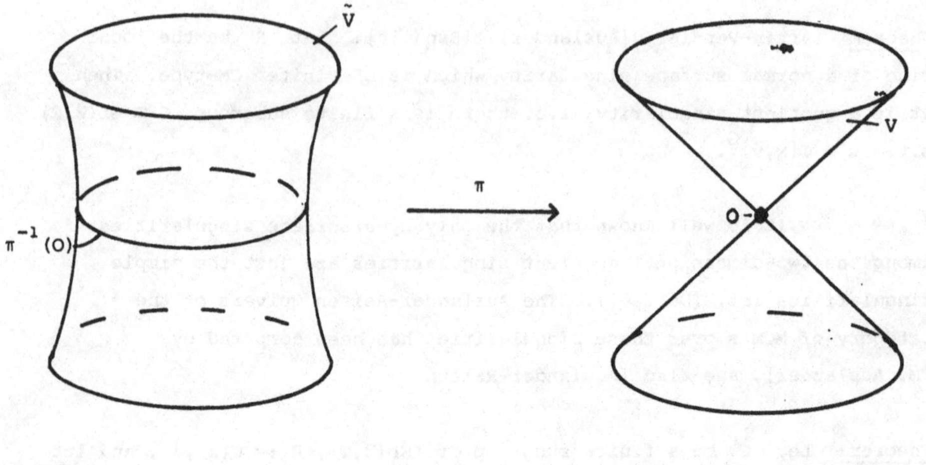

Figure 1

Such desingularisations always exist by Hironaka's theorem. In dimension two there is a unique "minimal" desingularisation, and there is an "algorithm" for computing this minimal desingularisation (cf. [Laufer]). For the minimal desingularisation of a suitable representative of a Kleinian singularity $\pi^{-1}(O)$ is a union of rational curves E_i. One associates to this configuration of curves a graph (the resolution graph) which has a vertex for each irreducible component E_i of $\pi^{-1}(O)$, and where two vertices are connected if and only if the corresponding curves intersect.

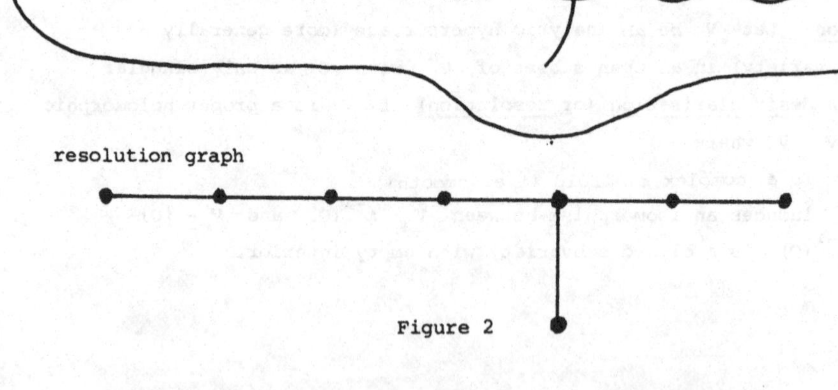

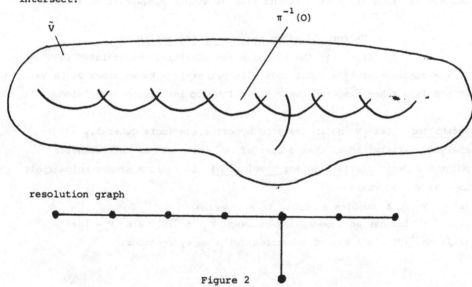

Figure 2

It turns out that the resoluti[illegible]phs of the Kleinian singularities are just the Dynkin diagrams of type A,D,E (the diagram associated to each such singularity is labelled in the list of (1.3) above). The occurence of these Dynkin diagrams both in the representation theory of finite subgroups of $SL(2,\mathbb{C})$ and in the desingularisation of the corresponding quotient singularity was one of the major motivations for the study of MCM's; [Gonzalez-Sprinberg and Verdier] found a construction which gives a link between these two types of diagrams:

Let V be a suitable representative of a Kleinian singularity and $\pi : \tilde{V} \to V$ the minimal desingularisation. If M is MCM over $\mathcal{O}_{(V,O)}$ we denote by $\mathcal{M}$ the corresponding sheaf of MCM's over V, it is locally free on $V-\{O\}$. As $\pi:\tilde{V}-\pi^{-1}(O) \to V-\{O\}$ is an isomorphism we get a locally free sheaf on $\tilde{V}-\pi^{-1}(O)$. Gonzalez-Sprinberg and Verdier construct a particular extension $\check{\mathcal{M}}$ of this sheaf over $\pi^{-1}(O)$, namely

$$\check{\mathcal{M}} := \pi^*(\mathcal{M})/\text{torsion}$$

The following theorem is due to [Gonzalez-Sprinberg, Verdier] (for other proofs see [Knörrer (1982)] and [Artin-Verdier] [Esnault-Knörrer]). The original formulation in the paper of Gonzalez-Sprinberg Verdier is in terms of McKay's tensor product formula; this description then follows from Auslander's work quoted above.

<u>Theorem</u>: (i) If M is MCM over $\mathcal{O}_{(V,O)}$ then the sheaf $\check{\mathcal{M}}$ is locally free on $\tilde{V}$.
(ii) There is a bijection $M_i \leftrightarrow E_i$ between the set of isomorphism classes of non-trivial indecomposable MCM's over $\mathcal{O}_{(V,O)}$ and the set of components of $\pi^{-1}(O)$ such that $\check{\mathcal{M}}_i\big|_{E_j}$ is non-trivial if and only if $i = j$.
(iii) There is an irreducible morphism $M_i \to M_j$ if and only if E_i and E_j intersect.

So the bijection of (ii) yields an isomorphism between the stable Auslander-Reiten quiver of MCM's on $\mathcal{O}_{(V,O)}$ and the resolution graph of the singularity (V,O).

Remark: The desingularisation of the Kleinian singularities can also be described in terms of the associated simple algebraic group (see [Slodowy]); however a description of the MCM's in terms of this algebraic group is still missing.

2.4 Simple Plane Curve Singularities:

As well as for the normal surface singularities one has a complete classification of all reduced curve singularities that are of finite CM-type.

Theorem ([Greuel-Knörrer]): Let R be the local ring of an isolated curve singularity. Then the following statements are equivalent:

(i) R is of finite CM-type

(ii) There exists the local ring S of a simple plane curve singularity and an embedding of rings $j : R \hookrightarrow \bar{S}$ of R into the normalisation $\bar{S}$ of S such that $S \subset j(R) \subset \bar{S}$.

Again the only hypersurface singularities among them are the simple one-dimensional hypersurface singularities. The Auslander-Reiten quiver of MCM's over simple plane curve singularities has been computed by [Dieterich-Wiedemann], a geometric interpretation of these quivers using group representations is given in [Knörrer].

3. HYPERSURFACE SINGULARITIES OF FINITE COHEN-MACAULAY TYPE

In this chapter we sketch the proof of the theorem stated in the introduction. M. Auslander has shown that a singularity of finite CM-type is necessary isolated, so - by the results mentioned in ch.2 - it suffices to deal with isolated hypersurface singularities of dimension ≥ 3. Before doing this we recall a general method for describing MCM's over hypersurface rings which is due to [Eisenbud]:

3.1 Matrix Factorisations:

We consider the local ring R of a hypersurface singularity, i.e. $R = P/_{(f)}$ where $P = \mathbb{C}\{x_1,\dots,x_n\}$, $f \neq 0$ in the maximal ideal $\mathfrak{m}_P$ of P. If M is MCM over R then - by a theorem of Auslander - Buchsbaum - M has a resolution of length one over P:

$$0 \to P^r \xrightarrow{\phi} P^r \to M \to 0 \quad \text{for some} \quad r \geq 1$$

As $f \,.\, P^r \subset \operatorname{im} \phi$ and ϕ is injective there is a morphism $\psi : P^r \to P^r$ such that

$$\phi \circ \psi = f \,.\, \mathrm{id} \qquad \text{and} \quad \psi \circ \phi = f \,.\, \mathrm{id}$$

<u>Definition</u>: A pair $(\phi : F \to G,\ \psi : G \to F)$ of morphisms between free P-modules, F, G with $\phi \circ \psi = f \,.\, \mathrm{id}$, $\psi \circ \phi = f \,.\, \mathrm{id}$ is called a <u>matrix factorisation</u> of f.

We have just seen that for each maximal CM module M over R there is a matrix factorisation (ϕ,ψ) of f such that $M \cong \operatorname{cok}\phi$. Conversely for any matrix factorisation (ϕ,ψ) of f the cokernel of ϕ is zero or an MCM over R.

If $(\phi : F \to G,\ \psi : G \to F)$ and $(\phi' : F' \to G',\ \psi' : G' \to F')$ are matrix factorisations of F then a pair of morphisms $(\alpha : F \to F',\ \beta : G \to G')$ is called a morphism of matrix factorisations between (ϕ,ψ) and (ϕ',ψ') if the following diagram commutes:

$$\begin{array}{ccccc} F & \xrightarrow{\phi} & F' & \xrightarrow{\psi} & F \\ \downarrow{\scriptstyle\alpha} & & \downarrow{\scriptstyle\beta} & & \downarrow{\scriptstyle\alpha} \\ F' & \xrightarrow{\phi'} & G' & \xrightarrow{\psi'} & F' \end{array}$$

In this way the matrix factorisations of f form an additive category, which we denote by $\mathcal{MF}(f)$. $(\phi,\psi) \longrightarrow \operatorname{cok}\phi$ is then a functor from $\mathcal{MF}(f)$ to the category of MCM's over R. The results above can be rephrased as follows:

Theorem ([Eisenbud]): The functor $(\phi,\psi) \longrightarrow \operatorname{cok}\phi$ from $\mathcal{MF}(f)$ to the category of MCM's over R induces a bijection between isomorphism classes of indecomposable objects in $\mathcal{MF}(f)$ and isomorphism classes of indecomposable MCM's over R (including the zero-module).

3.2 Double branched Covers:

To prove that all simple hypersurface singularities are of finite CM-type it suffices (by (1.3) and ch.2) to show

Theorem ([Knörrer]): Let $P = \mathbb{C}\{x_1,\dots,x_n\}$, $0 \neq f \in \mathfrak{m}_P$, $R := P/(f)$, $R_1 := P[y]/(f-y^2)$. Then R is of finite CM-type if and only if R_1 is of finite CM-type.

Outline of the proof: Let σ be the involution of R_1 mapping (x,y) to $(x,-y)$. We denote by $R_1[\sigma]$ the associated twisted group ring and by $CM_\sigma(R_1)$ the category of MCM's over $R_1[\sigma]$.
If $M \in CM_\sigma(R_1)$ then

$$M^\sigma := \{m \in M/\sigma(m) = m\} \text{ and}$$
$$M^a := \{m \in M/\sigma(m) = -m\}$$

and free modules over the invariant ring $R_1^\sigma \cong P$. Furthermore multiplication with y maps M^σ to M^a and M^a to M^σ. So we get a pair of P-linear maps

$$\phi : M^\sigma \to M^a, \qquad \psi : M^a \to M^\sigma$$

such that $\phi \circ \psi = f \,.\, \mathrm{id}$, $\psi \circ \phi = f \,.\, \mathrm{id}$. This construction yields a functor $\mathcal{A} : CM_\sigma(R_1) \to \mathcal{MF}(f)$, and one can show that $\mathcal{A}$ is an equivalence of categories. So R is of finite CM-type if and only if there are only finitely many isomorphism classes of indecomposable objects in $CM_\sigma(R_1)$. A standard argument on twisted group rings (cf. [Reiten-Riedtmann] 3.8) shows that this is the case if and only if R_1 is of finite CM-type.

If one adds two squares to the equation of an isolated hypersurface singularity one has the following stronger result:

<u>Theorem</u> ([Knörrer]): If $(V(f),0)$ is an isolated hypersurface singularity then the stable Auslander-Reiten quivers of MCM's over $R = P/(f)$ and $R_2 := P\{y,z\}/(f-y^2-z^2)$ are isomorphic.

We do not give a proof of this theorem here, but we describe the functor $\mathcal{MF}(f) \to \mathcal{MF}(f-y^2-z^2)$ that gives this isomorphism: Put $u: = y+iz$, $v: = u-iz$, then $y^2+z^2 = uv$. If $(\phi : P^r \to P^r, \psi : P^r \to P^r)$ is a matrix factorisation of f then

$$\left(\begin{pmatrix}\phi & u\\ v & \psi\end{pmatrix} : P_2^r \oplus P_2^r \to P_2^r \oplus P_2^r,\ \begin{pmatrix}\psi & -u\\ -v & \phi\end{pmatrix} : P_2^r \oplus P_2^r \to P_2^r \oplus P_2^r\right)$$

where $P_2 = P\{y,z\}$, is a matrix factorisation of $f-uv$. We will use this construction later in the following special case:

<u>Remark</u>: Let $Q_m(u_1,\ldots,u_m, v_1,\ldots, v_m) = u_1v_1 +\ldots+ u_mv_m$. Then there exists a matrix factorisation $(A_m(u,v), B_m(u,v))$ of Q_m consisting of $2^{m-1} \times 2^{m-1}$ matrices A_m, B_m whose entries are all either 0 or $\pm u_i$, $\pm v_i$.

3.3 <u>Constructing Families of MCM's</u>:

We now report on the construction of [Buchweitz-Greuel-Schreyer] which shows that there are infinitely many isomorphism classes of indecomposable MCM's over non-simple isolated hypersurface singularity. For this purpose one has to construct families of MCM's over the singularities in question and to show that the elements of these families are pairwise non-isomorphic. To distinguish isomorphism classes of MCM's they use the following invariant:

<u>Definition</u>: Let $(\phi : P^r \to P^r, \psi : P^r \to P^r)$ be a matrix factorisation of f. The <u>Fitting ideal</u> of (ϕ,ψ) is the ideal in P generated by the entries of the matrices ϕ and ψ. Obviously it only depends on the isomorphism class of the matrix factorisation (ϕ,ψ), so it is an invariant of the maximal CM module $\mathrm{cok}\,\phi$.

The construction of families of MCM is based on

<u>Lemma</u>: Let $\underline{g} = (g_1,\ldots,g_n)$ be a sequence in $\mathfrak{m}_P$ such that f lies in the square of the ideal $\langle\underline{g}\rangle$ generated by $g_1,\ldots,g_n$. Then there is an MCM over R whose Fitting ideal is equal to $\langle\underline{g}\rangle$.

<u>Proof</u>: Write $f = \sum_{i=1}^{n} g_i h_i$ with $h_i \in \langle\underline{g}\rangle$. Let M be the MCM defined by the matrix factorisation $(A_n(g,h), B_n(g,h))$ that is obtained from the matrix factorisation $(A_n(u,v), B_n(u,v))$ of Q_n as in (3.2) by substituting g_i for u_i and h_i for v_i. The Fitting ideal of this matrix factorisation is generated by the g_i and h_i, so it is equal to $\langle\underline{g}\rangle$.

The final step in the proof of [Buchweitz-Greuel-Schreyer] is to show that non-simple isolated hypersurface singularities can in fact be characterised by the property that there are infinitely many sequences $\underline{g}$ as in the lemma above whose ideals $\langle\underline{g}\rangle$ are all different.

4. Miscellaneous Remarks

4.1 [Herzog] has shown that Gorenstein singularities which are of finite CM-type are actually hypersurface singularities. So the theorem stated in the introduction characterises the simple hypersurface singularities among all Gorenstein singularities.

4.2 Apart from the simple singularities the only other isolated hypersurface singularity for which all MCM's have been classified (till August 85) is the unimodular singularity

$$\tilde{E}_8 : y(y+x^2)(y-x^2) + z_3^2 + \ldots + z_n^2 = 0$$

Its Auslander-Reiten quiver has been computed by E. Dieterich (Bielefeld). It turns out that this singularity is of tame CM-type, i.e. there are countably many one-parameter families of MCM's that cover all isomorphism classes of indecomposable MCM's.

A.Schappert (Kaiserslautern) showed all MCM's of rank one over the unimodular curve singularities can be covered by finitely many one-parameter families.

4.3 The fact that simple hypersurface singularisation can be characterised by a finiteness property both with respect to their deformation theory and to their representation theory raises the question whether the number of parameters of MCM's (with certain discrete invariants, e.g. rank) over a (hypersurface) - singularity is upper semicontinuous under deformation. For the case of finite dimensional algebras see [Gabriel].

4.4 For plane curve singularities all MCM's of rank r can be obtained as a specialisation of the free module of rank r (cf. [Rego]). It would be interesting to get some more insight in the specialisation of MCM's over simple curve - and surface singularities.

4.5 From results of [Bruns] it follows that a non-trivial MCM over the local ring of an n-dimensional isolated hypersurface has at least rank $\frac{1}{2}n-1$. It is not known whether this bound is sharp.

4.6 R. Buchweitz has show that the category of MCM's over a Gorenstein ring R (modulo morphism factoring over projectives) is isomorphic to the quotient of the derived category of R-modules with bounded cohomology of finite type by the subcategory of perfect complexes.

4.7 If M is MCM over a two-dimensional quotient singularity (V,O) and $\mathcal{M}$ is the corresponding sheaf on V as in (2.3) then $\mathcal{M}|_{V-\{O\}}$ admits an integrable connection (coming from the trivial connection on the pull back $\tau^*(\mathcal{M}) \cong \mathcal{O}^r_{\mathbb{C}^2,O}$ of $\mathcal{M}$ under the quotient map $\tau : (\mathbb{C}^2,O) \to (V,O)$). It would be interesting to know which MCM's on a normal surface singularity in general admit an integrable connection on the complement of the singular point, i.e. "come from" a representation of the local fundamental group.

REFERENCES

Arnold, V.I. (1981). Singularity Theory - Selected papers. London Math. Soc. Lecture Note Series 53, Cambridge University Press.

Arnold, V.I., Gusein-Zade, S.M. & Varchenko, A.N. (1985). Singularities of Differentiable Maps I. Birkhäuser.

Artin, M. & Verdier, J-L. (1985). Reflexive modules over rational double points. Math. Ann. 270, 79-82.

Auslander, M. Almost split sequences and rational singularities. Trans. AMS.

Auslander, M. & Reiten, I. McKay quivers and extended Dynkin diagrams. Preprint.

Bruns, W. (1981). The Eisenbud-Evans generalised principal ideal theorem and determinantal ideals. Proc. AMS 83, 19-26.

Buchweitz, R., Greuel, G.M. & Schreyer, F.-O. Maximal Cohen-Macaulay modules on hypersurface singularities II. In preparation.

Damon, J. (1984). The Unfolding and Determinary Theorems for Subgroups of A and K. Memoirs AMS 50.

Dieterich, E. & Wiedemann, A. The Auslander-Reiten quiver of a simple curve singularity. Trans. AMS.

Durfee, A. (1979). Fifteen characterisations of rational double points and simple critical points. L'Enseignement Math. 25, 131-163.

Eisenbud, D. (1980). Homological Algebra on a complete intersection, with an application to group representations. Trans. AMS 260, 35-64.

Esnault, H. Reflexive modules on quotient singularities. J. Reine Angew. Math.

Esnault, H. & Knörrer, H. (1985). Reflexive modules over rational double points. Math. Ann. 272, 545-548.

Gabriel, P. (1975). Finite representation type is open. In : Representation of Algebras (Ed. Dlab, Gabriel), pp.132-155. Lecture Notes in Math. 488, Springer.

Gonzalez-Sprinberg, G. & Verdier, J.-L. (1983). Construction geometrique de la correspondance de McKay. Ann. Sc. Ecole. Norm. Sup. 16, 409-449.

Greuel, G.-M. & Knörrer, H. (1985). Einfache Kurvensingularitäten und torsionsfreie Moduln. Math. Ann. 270, 417-425.

Gunning, R. & Rossi, H. (1965). Analytic Functions of Several Complex Variables. Prentice-Hall, Englewood Cliffs, N.J.

Herzog, J. (1978). Ringe mit nur endlich vielen Isomorphishlassen von maximalen unzerlegbaren Cohen-Macaulay-Moduln. Math. Ann. 233, 21-34.

Knörrer, H. (1982). Group representations and resolution of rational double points. To appear in Proc. Conf. Group Theory, Montreal.

Knörrer, H. Maximal Cohen-Macaulay modules on hypersurface singularities I. In preparation.

Laufer, H. (1971). Normal Two-Dimensional Singularities.Annals Math. Studies 71, Princeton University Press.

Looijenga, E. (1984). Isolated Singular Points on Complete Intersections. London Math. Soc. Lecture Note Series 77, Cambridge University Press.

McKay, J. (1980). Graphs, singularities and finite groups. Proc. Symp. Pure Math. 37, 183-186.

Newstead, P. (1978). Introduction to Moduli Problems and Orbit Spaces. Springer.

Rego, C. (1982). Compactification of the space of vector bundles on a singular curve. Comment. Math. Helvetici 57, 226-236.

Reiten, I. & Riedtmann, Chr. (1985). Skew group algebras in the representation theory of Artin algebras. J. of Algebra 92, 224-282.

Slodowy, P. (1983). Platonic Solids, Kleinian Singularities, and Lie groups In Algebraic Geometry (Ed. Dolgachev, I.), pp.102-138. Lecture Notes in Math, 1008, Springer.

Steinberg, R. Subgroups of SU_2 and Dynkin diagrams. Preprint.

Wall, C.T.C. (1984). Notes on the classification of singularities. Proc. London Math. Soc. 48, 461-513.

Almost Split Sequences and Algebraic Geometry

Maurice Auslander
Brandeis University
Waltham, MA 02254 U.S.A.

Our main purpose in this paper is to explore some instances of almost split sequences occuring in algebraic geometry. We will be dealing mainly with three topics:

(1) The connections between the structure of the almost split sequences for reflexive modules over rational double points and the desingularization graph of the singularity;

(2) The existence of almost split sequences for coherent sheaves over nonsingular and Gorenstein projective curves;

(3) The question of which complete integrally closed Cohen-Macaulay local domains are of finite Cohen-Macaulay type, i.e. have, up to isomorphism only a finite number of indecomposable Cohen-Macaulay modules.

This paper is almost a verbatim account of the last two lectures I gave at the Durham symposium on the representation theory of algebras. Since these lectures were purely expository, no proofs are given.

§1. Complete rational double points.

Throughout this section k is an algebraically closed field. We recall that the complete rational double points over k can be described as follows. Let k[[u,v,w]] be the ring of formal power series in three variables u, v, w. Then the complete rational double points over k are k-algebras R of the form k[[u,v,w]]/(f(u,v,w)) with f(u,v,w) a non-zero element in the maximal ideal (u,v,w) of k[[u,v,w]] such that R is an integrally closed domain with a finite class group. Although all the f(u,v,w) in (u,v,w) which define the rational double points, up to isomorphism, are known explicitly, we do not give the list here since we will be mainly interested in those complete rational double points which are quotient singularities. A description of these types of complete rational singularities can be given as follows.

Let G be a finite subgroup of SL(2,k) with order relatively prime to the characteristic of k and let S = k[[x,y]]. Then the representation given by the inclusion $G \hookrightarrow SL(2,k)$ gives an action of G on the two dimensional k-vector space of S generated by X and Y. This in turn gives a linear k-algebra action of G on S. Then $R = S^G$, the k-subalgebra of S consisting of s in S such that $\sigma(s) = s$ for all σ in G, is a complete rational double point. In characteristic zero one obtains all the complete rational double points in this way.

While we will be interested mainly in the structure of almost split sequences in the category of reflexive modules over complete rational double points, we begin by first considering almost split sequences in the more general case of arbitrary complete integrally closed local noetherian domains R of Krull dimension 2. Such an R is an order in the sense described in [3] and [2] with lattices the reflexive R-modules. Therefore the category of reflexive R-modules has almost split sequences. For these types of rings there is a particularly interesting way of constructing at least some of the almost split sequences, as we now explain.

There is a uniquely determined reflexive R-module ω of rank one, called the dualizing module, such that $Ext^i_R(R/\underline{m}, \omega)$ is zero for i < 2 and is $R/\underline{m}$ for i=2. Then any two exact sequences of the form $0 \to \omega \xrightarrow{g} E \xrightarrow{f} R \to R/\underline{m} \to 0$ representing non-zero elements of $Ext^2_R(R/\underline{m}, \omega)$ are isomorphic and these uniquely determined exact sequences have the following properties:

(a) E is reflexive

(b) $f: E \to R$ is right almost split in the category of reflexive R-modules (notation: Ref R), i.e. any $g: X \to R$ in Ref R which is not a splittable surjection can be lifted to E.

(c) $g: \omega \to E$ is left almost split in Ref R, i.e. any $h: \omega \to Y$ in Ref R which is not a splittable injection can be extended to Y.

This uniquely determined exact sequence $0 \to \omega \to E \to R \to R/\underline{m} \to 0$ is called the fundamental exact sequence for reasons which will become clear shortly. The fundamental sequence will always be denoted by $0 \to \omega \to E \to R \to R/\underline{m} \to 0$. It should be noted that the properties listed above actually determine the fundamental exact sequence. For if $0 \to A \xrightarrow{u} B \xrightarrow{v} R$ is an exact sequence in Ref R such that A is indecomposable, u is left almost split and v is right almost split, then Im $v = \underline{m}$ and the exact sequence $0 \to A \to B \to R \to R/\underline{m} \to 0$ is isomorphic to the fundamental exact sequence. We now describe the connection between the fundamental exact sequence and almost split sequences in Ref R. To this end, it is convenient to have the following notion.

If A and B are in Ref R, then $A \otimes_R B$ is not generally in Ref R. However, $\mathrm{Hom}_R(\mathrm{Hom}_R(A \otimes_R B, R), R)$, which we denote by $A \cdot B$, is in Ref R. Clearly each morphism $f: B \to C$ in Ref R induces in an obvious way a morphism $A \cdot B \to A \cdot C$. Moreover the functor $A \cdot B$ has the property that for all A, B, C in Ref R we have isomorphisms in Ref R

$$\mathrm{Hom}_R(A \cdot B, C) \to \mathrm{Hom}_R(A, \mathrm{Hom}_R(B,C))$$

which are functorial in A, B, C. So $A \cdot B$ is the "tensor product" in the category Ref R. In particular, $A \cdot R$ is naturally isomorphic to A.

Now returning to our considerations of the fundamental exact sequence we have the following important property (see [5]).

<u>Proposition 1.1</u> Let $0 \to \omega \to E \to R \to R/\underline{m} \to 0$ be the fundamenatal exact sequence. For each indecomposable nonprojective A in R we have the following:

(a) $0 \to A \cdot \omega \to A \cdot E \to A \cdot R = A \to 0$ is exact.

(b) The exact sequence $0 \to A \cdot \omega \to A \cdot E \to A \to 0$ is either almost split or splits.

(c) $0 \to A \cdot \omega \to A \cdot E \to A \to 0$ is almost split if rank A is not divisible by the characteristic of $R/\underline{m}$.

(d) If $R/\underline{m}$ is algebraically closed, then $0 \to A \cdot \omega \to A \cdot E \to A \to 0$ is almost split if and only if rank A is not divisible by characteristic of $R/\underline{m}$.

Therefore we see that when $R/\underline{m}$ is of characteristic zero, to describe the modules in the almost split sequences, it suffices to know the fundamental exact sequence $0 \to \omega \to E \to R \to R/\underline{m} \to 0$ and what $\omega \cdot A$ and $E \cdot A$ are for nonprojective indecomposable A. It should be remarked that if $0 \to A \to B \to C \to 0$ is an almost split sequence in Ref R, then $A = C \cdot \omega$ even if $0 \to \omega \cdot C \to E \cdot C \to C \to 0$ splits. Since R is Gorenstein if and only if $\omega = R$, we have that for a Gorenstein ring the almost split sequences all have the form $0 \to C \to B \to C \to 0$, i.e. have the same ends. In particular, this is the case for complete rational double points since they are hypersurfaces.

We end this general discussion by recalling the definition of the AR quiver of R in the case $R/\underline{m}$ is algebraically closed. The AR quiver of R is a directed graph whose vertices are the isomorphism classes of indecomposable modules in Ref R and whose arrows are described as follows. Let L_1, L_2 be two indecomposable reflexive R-modules. Suppose $L_1 = R$. Then the number of arrows from L_2 to L_1 is the multiplicity of L_2 in E, i.e. the number of copies of L_2 occuring in the representation of E as a sum of indecomposable modules, where E is the same module E occuring in the fundamental exact sequence. If L_1 is not projective, then there is always an almost split sequence $0 \to \omega \cdot L_1 \to B \to L_1 \to 0$ and the number of arrows from L_2 to L_1 is the multiplicity of L_2 in B.

It is obvious that there is only a finite number of arrows coming into and going out of any vertex in the AR quiver of R. Also if R is Gorenstein, it follows from our previous remarks concerning the structure of the fundamental exact sequence and almost split sequences, the number of arrows from vertex A to vertex B is the same as the number of arrows going from B to A.

With these preliminary generalities in mind we return to the case R is a complete rational double point. Our initial purpose is to discuss the connections between the AR quiver of R and the minimal desingularization graph of R. I briefly recall how the desingularization graph is constructed (see Knörrer's article in these proceedings for a fuller discussion). The vertices of the graph are the irreducible components $E_1,\dots,E_t$ of the exceptional divisor of the minimal resolution of singularities and one draws an edge from E_i to E_j $(i \neq j)$ if $E_i \cap E_j \neq \emptyset$. It is an old result that the resolution graphs of rational double points are the Dynkin diagrams of types A, D and E in all characteristics.

Motivated by work of McKay, which will be discussed later, Gonzalez-Sprinberg and Verdier [10] and Knörrer (unpublished) described a bijection between the isomorphism classes of nonprojective indecomposable reflexive R-modules and the exceptional divisors when R is a quotient singularity. Artin and Verdier [1], using quite different techniques, generalized this result to arbitrary complete rational double points. Amongst other things, this shows that there is only a finite number of nonisomorphic indecomposable reflexive R-modules when R is an arbitrary complete rational double point. Since this result gives a bijection between the vertices of the AR quiver of R with vertex corresponding to R deleted and the vertices of the desingularization graph of R, it is natural to wonder if the underlying graph of the AR quiver of R with R deleted is isomorphic to the desingularization graph of R.

Since the desingularization graph of R is always a Dynkin diagram, it is natural to ask if the underlying graph of the AR quiver with R deleted is a Dynkin diagram. This was shown by Auslander and Reiten, who showed that the AR quiver for R is an extended Dynkin diagram with arrows in both directions and that one gets a Dynkin diagram with arrows in both directions by deleting the vertex corresponding to R. The proof is purely combinatorial and depends in an essential way on the fact that the fundamental exact sequence has combinatorial properties similar to those of almost split sequence. It then follows from work of Ésnault and Knörrer [9], simplifying work of Gonzalez-Sprinberg and Verdier, that we have the following.

Theorem 1.2 Let R be a complete rational double point. Then the Artin-Verdier correspondence between isomorphism classes of nonprojective indecomposable reflexive R-modules and the exceptional divisors in the desingularization of R extends to an isomorphism between the desingularization graph of R and the underlying graph of the stable AR quiver of R (i.e. the quiver obtained from the AR quiver by leaving out R).

In the case the complete rational double point R is a quotient singularity, there is a group representation way of constructing a graph naturally isomorphic to the desingularization graph of R. This fact was first shown by McKay in a somewhat empirical way and it was a desire to find a more organic explanation of McKay's result that motivated the work of Gonzalez-Springberg and Verdier and the independent work of Knörrer cited earlier. We end our discussion of complete rational double points with a description of McKay's result.

We begin with the notion of the McKay quiver of a group representation. Let G be a finite group whose order is relatively prime to the characteristic of the algebraically closed field k. Then k[G], the group ring of G over k, is semisimple and let $V_1,\dots,V_t$ be a complete set of nonisomorphic simple k[G]-modules. Then associated with each k[G]-module V is a quiver called the McKay quiver of V which we now describe. The vertices of the McKay quiver of V are the isomorphism classes $[V_1],\dots,[V_t]$ of simple k[G]-modules and the number of arrows from $[V_i]$ to $[V_j]$ is given by the multiplicity of V_i in $V \otimes_k V_j$ for all i and j.

Now suppose $G \subset GL(2,k)$ is a finite subgroup whose order is relatively prime to the characteristic of k and not containing any pseudoreflections. Then associated with the inclusion $G \hookrightarrow GL(2,k)$ is a two-dimensional k[G]-module V as well as a linear action of G on the k-algebra $S = k[[X,Y]]$. Let $R = S^G$. Then R is a local complete integrally closed noetherian domain of dimension 2, so it has an AR-quiver. Then we have the following result due to Auslander [4].

Theorem 1.3 Under the above hypothesis we have that the AR quiver of R is canonically isomorphic to the McKay quiver of V and under this isomorphism the vertex corresponding to R is carried to the vertex corresponding to the trivial representation of G.

As a consequence of Theorems 1.2 and 1.3 we have obtained McKay's result.

Theorem 1.4 Let $G \subset SL(2,k)$ be a finite subgroup of order relatively prime to order of characteristic k. Let V be the two-dimensional k[G]-module corresponding to $G \hookrightarrow SL(2,k)$ and let R be the complete rational double point which is the fixed point ring of the corresponding linear action of G on $k[[X,Y]]$. Then the underlying graph of the McKay quiver of V with the trivial representation deleted is isomorphic to the desingularization graph of R.

§2. Projective curves

Our purpose in this section is to point out how the existence theorems for almost split sequences for graded modules over graded algebras discussed in our previous paper in these proceedings give the existence of almost split sequences for some subcategories of locally free coherent sheaves over certain types of projective Gorenstein varieties. As a consequence of these general results one gets the existence of almost split sequences for locally free sheaves over Gorenstein curves and the existence of almost split sequences for the category of coherent sheaves over nonsingular curves. These results for curves were obtained independently by A. Schofield for algebraically closed fields using sheaf theoretic methods directly.

Let k be an infinite field, not necessarily algebraically closed. Suppose X is a connected projective k-variety of dimension $d \geq 1$ which is Gorenstein, i.e. $\mathcal{O}_{X,x}$ is Gorenstein for all x in X. Suppose further that X can be embedded in a projective space in such a way that its homogeneous coordinate ring S is a graded Cohen-Macaulay ring. Now the "sheafification" functor $M \to \tilde{M}$ from the category of finitely generated graded S-modules with degree zero morphisms to Coh(X), the category of coherent sheaves on X, induces an equivalence between the category of reflexive graded S-modules and the category of reflexive coherent sheaves on X.

The fact that X is Gorenstein is equivalent to $S_{\underline{p}}$ being Gorenstein for all homogeneous prime ideals $\underline{p}$ different from the unique maximal homogeneous ideal m. Therefore S is a graded order in the terminology of [2] and [5]. Also the S-lattices are the graded Cohen-Macaulay modules L such that $L_{\underline{p}}$ is $S_{\underline{p}}$ free for all homogeneous prime ideals $p \neq m$. Therefore for each L in Latt S, the category of S-lattices, we have that $\tilde{L}$ is a locally free coherent sheaf on X. Since the S-lattices are reflexive S-modules, it follows that Latt S is equivalent to a certain subcategory of locally free coherent sheaves on X. Hence the fact that the category Latt S has almost split sequences implies that the corresponding subcategory of locally free sheaves on X also has almost split sequences. Unfortunately if the dimension of X is greater than 1, the category of locally free sheaves on X corresponding to Latt S is not the whole category of locally free sheaves on X, since in this case the category of locally free sheaves on X does not have almost split sequences. However the situation is quite different when the dimension of X is one. We therefore denote the rest of this section to the case X is a curve.

Let X be a connected Gorenstein projective curve. Then one can always find an embedding $X \hookrightarrow \mathbf{P}^n$ such that the homogeneous coordinate ring S of X is Cohen-Macaulay. Moreover for each locally free sheaf F on X there is a lattice L on S such that $\tilde{L} \cong F$. Therefore the category of locally free sheaves on X is equivalent to the category Latt S. This gives the following result [5].

<u>Theorem</u> 2.1 Let X be a connected Gorenstein projective curve. Then the category of locally free sheaves on X has almost split sequences, i.e. for each indecomposable locally free sheaf f, there exist almost split sequences $0 \to F \to G \to H \to 0$ and $0 \to U \to V \to F \to 0$.

If X is nonsingular, then we have the following somewhat stronger result.

<u>Theorem</u> 2.2 Let X be a connected nonsingular projective curve. Then the category of all coherent sheaves over X has almost split sequences.

In fact, one can show that an irreducible projective variety X has the property that the category of coherent sheaves on X has almost split sequences if and only if X is a nonsingular curve.

We end this section by pointing out that for connected Gorenstein projective curves, there is a fundamental sequence with properties analogous to those of the fundamental sequence for local complete integrally closed noetherian domains of dimension 2 discussed in section one.

Let X be a connected Gorenstein projective curve. Then there is a unique nonsplit exact sequence (up to isomorphism) of coherent sheaves $0 \to \omega \to E_{\mathcal{O}_X} \to \mathcal{O}_X \to 0$ where ω is the dualizing sheaf of X which we call the fundamental sequence for X.

Proposition 2.3 Let X be a connected Gorenstein projective curve over an algebraically closed field k and let $0 \to \omega \to E_{\mathcal{O}_X} \to \mathcal{O}_X \to 0$ be the fundamental sequence of X. Then for each locally free indecomposable sheaf F we have that the sequence $0 \to \omega \otimes_{\mathcal{O}_X} F \to E_{\mathcal{O}_X} \otimes_{\mathcal{O}_X} F \to \mathcal{O}_X \otimes_{\mathcal{O}_X} F = F \to 0$ is exact and is an almost split sequence if ch k does not divide rank F, otherwise it is split.

Let X be an irreducible nonsingular projective curve of genus 1 over an algebraically closed field of characteristic 0, then the almost split sequences for locally free indecomposable sheaves on X can be explicitly described using Atiyah's classification of locally free sheaves on X together with the method of constructing almost split sequences given by Proposition 2.3.

Another case when the almost split sequences of locally free sheaves can be described explicitly is when $X = \mathbf{P}^1$ over an arbitrary field k.

§3. Finite Cohen-Macaulay type

For the sake of simplicity we assume throughout this section that our ground field k is **C**, the complex numbers, even though this is sometimes not really necessary. Unless stated to the contrary, by a **C**-algebra R we mean a factor ring of a formal power series ring $\mathbf{C}[[X_1,\ldots,X_n]]$ in a finite number of variables $X_1,\ldots,X_n$ over **C**, i.e. R is a **C**-algebra which is a complete local noetherian ring with residue field **C**. We will be mainly concerned in this section with the question of when a Cohen-Macaulay

$\mathbb{C}$-algebra is of finite Cohen-Macaulay type, i.e. has only a finite number of nonisomorphic indecomposable (maximal) Cohen-Macaulay modules.

While there are not many general facts known about Cohen-Macaulay $\mathbb{C}$-algebras of finite Cohen-Macaulay type, we do have the following due to Auslander [4].

Theorem 3.1   Suppose R is a Cohen-Macaulay $\mathbb{C}$-algebra of finite Cohen-Macaulay type. Then

a) R is an isolated singularity, i.e. $R_{\underline{p}}$ is regular for each nonmaximal prime ideal $\underline{p}$ of R.

b) If dim R $\geq$ 2, then R is an integrally closed domain.

In the first section of this paper, it was pointed out that the complete rational double points are of finite Cohen-Macaulay type. Since these are all hypersurfaces, it is natural to ask which hypersurfaces R are of finite Cohen-Macaulay type. This question has been completely settled by Knörrer and Buchweitz-Greuel-Schreyer who have shown the following (see Knörrer's paper in these proceedings for an account of this work).

Theorem 3.2   A hypersurface R is of finite Cohen-Macaulay type if and only if R is a simple Arnol'd singularity.

Since the simple Arnol'd singularities are explicitly known, this result gives a complete classification of the hypersurfaces of finite Cohen-Macaulay type. So we will concentrate in the rest of this discussion on the problem of which isolated singularities which are not hypersurfaces are of finite Cohen-Macaulay type. Here the situation is not so satisfactory, since there is no complete classification of these singularities. What follows is an account of what is known about this question at the present time.

In view of the solution of this problem for hypersurfaces, it is natural to consider next complete intersections or, more generally,

Gorenstein singularities. This question was answered by Herzog [11], who showed the following somewhat surprising result.

Proposition 3.2 If R is a Gorenstein **C**-algebra of finite Cohen-Macaulay type, then R is a hypersurface.

In other words, the classification of hypersurfaces of finite Cohen-Macaulay type also gives more generally the classification of Gorenstein singularities of finite Cohen-Macaulay type.

Now let $G \subset GL(2,\mathbf{C})$ be a finite group containing no pseudoreflections. Then G acts linearly on $S = \mathbf{C}[[X,Y]]$. Implicit in the statement of Theorem 1.3 is the fact that $R = S^G$ is of finite Cohen-Macaulay type. This was originally shown by Herzog [11] who seems to have been the first to consider the question of when Cohen-Macaulay **C**-algebras are of finite Cohen-Macaulay type. Singularities of type $R = S^G$ are called quotient singularities. Since these quotient singularities have been classified for some time [8], one would have a complete classification of two-dimensional Cohen-Macaulay **C**-algebras of finite Cohen-Macaulay type if one knew that the quotient singularities are the only two-dimensional Cohen-Macaulay **C**-algebras of finite Cohen-Macaulay type. That this is indeed the case has been shown by Artin-Verdier, Auslander [5] and Ésnault, each using somewhat different techniques.

Having classified the Cohen-Macaulay **C**-algebras of dimension 2, we turn our attention to Cohen-Macaulay **C**-algebras R of dimension greater than 2. Here the situation is very unsatisfactory. There are only two Cohen-Macaulay **C**-algebras of dimension greater than 2 which are not hypersurfaces which are known to be of finite Cohen-Macaulay type. These are both of dimension three and were found by Auslander-Reiten. It is an obvious question if there are any others. We now describe these two **C**-algebras.

In view of the results concerning two-dimensional **C**-algebras, it is natural to ask in dimensions higher than two which quotient singularities if any are of finite Cohen-Macaulay type. This question is completely answered by the following result of Auslander-Reiten.

Theorem 3.3 Let $n \geq 3$ and let G be a finite subgroup of $GL(n,\mathbf{C})$ not containing any pseudoreflections. Then $R = \mathbf{C}[[X_1,\ldots,X_n]]^G$ is of finite Cohen-Macaulay type if and only if

a) $n=3$

b) $G = \mathbf{Z}/2\mathbf{Z}$

c) $\sigma(X_i) = -X_i$ for all $i=1,2,3$ where σ generates G.

Since the **C**-algebra described in Theorem 3.3 is not Gorenstein, it is not a hypersurface and so it gives one of our desired examples. To show that the **C**-algebra R described in Theorem 3.3 is of finite Cohen-Macaulay type can be done by using either techniques of homological algebra or by investigating the almost split sequences in the category of Cohen-Macaulay R-modules. The fact that no other quotient singularity has finite Cohen-Macaulay type follows from the following more general considerations.

Let S be a **C**-algebra which is a Cohen-Macaulay integrally closed domain of dimension at least 3. Let G be a finite group of **C**-algebra automorphisms of S. Then $R = S^G$ is also a **C**-algebra which is a Cohen-Macaulay integrally closed domain with dim S = dim R. Further assume that for each prime ideal $\underline{p}$ of R of height one, $S_{\underline{p}}$ is an unramified extension of $R_{\underline{p}}$. Finally assume that R is of finite Cohen-Macaulay type. Then we have the following consequences:

a) S is of finite Cohen-Macaulay type.

b) dim S = 3

c) $G \cong \mathbf{Z}/2\mathbf{Z}$ and the multiplicity of the trivial $\mathbf{C}[G]$-module **C** in the $\mathbf{C}[G]$-module $\underline{m}/\underline{m}^2$ is at most one where $\underline{m}$ is the maximal ideal of S.

The proof of this result depends heavily on some elementary facts concerning when finite dimensional algebras (not necessarily commutative) of radical square zero are of finite representation type.

Our final example came from a suggestion of Eisenbud that scrolls might give examples of Cohen-Macaulay **C**-algebras of finite Cohen-Macaulay type which are not hypersurface. This led to the following result of Auslander-Reiten.

Theorem 3.4 The only rational normal scroll R which is not a hypersurface and is of finite Cohen-Macaulay type is the scroll (2,1) which is the three dimensional **C**-algebra $\mathbf{C}[[X_0, X_1, X_2, Y_0, Y_1]]$ modulo the ideal generated by the 2×2 minors of the matrix

$$\begin{pmatrix} x_0, & x_1, & y_0 \\ x_1, & x_2, & y_1 \end{pmatrix}$$

The proof that the rational normal scroll (2,1) is of finite Cohen-Macaulay type depends on describing the almost split sequences in the category of Cohen-Macaulay R-modules which is known to have almost split sequences since scrolls are isolated singularities. The fact that the rational normal scroll (2,1) is the only rational normal scroll which is not a hypersurface and which is of finite Cohen-Macaulay type follows from an analysis of the extensions of certain rank 1 Cohen-Macaulay modules by other rank 1 Cohen-Macaulay modules.

As stated earlier, there are not very many special properties known of Cohen-Macaulay **C**-algebras R of finite Cohen-Macaulay type. One that seems to be of some interest is that $G_0(R)$, the Grothendieck group of R which is the free abelian group generated by all modules modulo short exact sequences, is a finitely generated abelian group. While $G_0(R)$ being finitely generated does not imply for a Cohen-Macaulay **C**-algebra R, that R is of finite Cohen-Macaulay type, it does put serious restrictions on R. In some sense, it can be viewed as a generalization of

R being a rational singularity, since if dim R = 2, then R is a rational singularity if and only if $G_0(R)$ is finitely generated. Therefore it seems reasonable to study Cohen-Macaulay **C**-algebras R with $G_0(R)$ finitely generated, even if R has infinite Cohen-Macaulay type. One source of examples of Cohen-Macaulay **C**-algebras with finitely generated Grothendieck groups are the quotient singularities as shown by Auslander-Reiten [7] most of which of course are not of finite Cohen-Macaulay type when the dimension is greater than 2 (see Theorem 3.3). It would be interesting to know that if S is a Cohen-Macaulay integrally closed **C**-algebra and G is a finite group of C-automorphisms of S whose fixed point ring R is unramified at height one primes in S, then does $G_0(S)$ being finitely generated imply $G_0(R)$ is finitely generated. If S is of finite Cohen-Macaulay type, the answer is yes, which generalizes the quotient singularity situation. However, nothing seems to be known in general.

References

1. M. Artin and J.-L. Verdier, "Reflexive modules over rational double points", Math. Ann. 270, 79-82 (1985).
2. M. Auslander, "A survey of existence theorems for almost split sequences" (these proceedings).
3. M. Auslander, "Functors and morphisms determined by objects", Proc. Conf. on Representation Theory (Philadelphia 1976). (Marcel Dekker, 1978), 1-244.
4. M. Auslander, "Isolated singularities and existence of almost split sequences"
5. M. Auslander, "Rational singularities and almost split sequences", Trans. Amer. Math. Soc. (to appear).
6. M. Auslander and I. Reiten, "Almost split sequences for graded modules I".
7. M. Auslander and I. Reiten, "Grothendieck groups of algebras and orders", J. of Pure and Applied Algebra, 39, 1-51 (1986).
8. E. Brieskorn, "Rationale singularitaten komplexer flachen " , Inventiones Math. 4, 336-358 (1968).
9. H. Ésnault and H. Knörrer, "Reflexive modules over rational double points", Math. Ann. 272, 545-548 (1985).
10. G. Gonzalez-Sprinberg and J.-L. Verdier, "Construction géométrique de la correspondence de McKay", Ann. Sci. Ec. Norm. Supér, IV, Sec. 16, 409-449 (1983).
11. J. Herzog, "Ringe mit nur endlich vielen isomorphieklassen von maximalen, unzerlegbaren Cohen-Macaulay-Moduln", Math. Ann. 233, 21-34 (1978).

REPRESENTATION RINGS OF FINITE GROUPS

David Benson
Mathematics Department, Northwestern University, Evanston, IL 60201, U.S.A.

Preface

At the Ottawa 1984 fourth international conference on the representations of algebras ("ICRA IV"), I gave a series of three expository lectures entitled "Modules for finite groups : representation rings, quivers and varieties". The main theme of those lectures was to demonstrate the connections depicted in the following diagram.

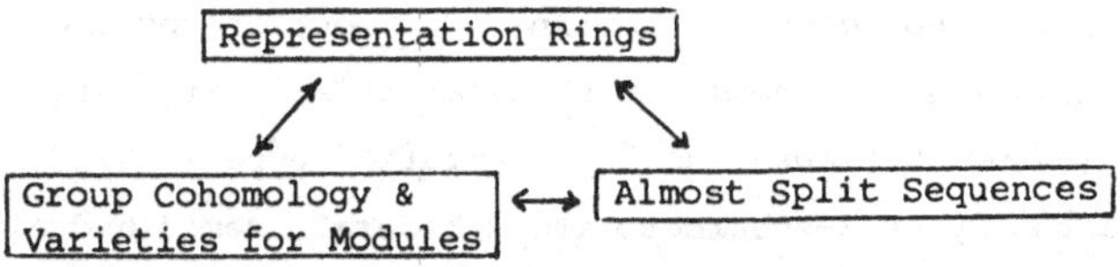

These lectures were written up, and will appear in the proceedings, published by Springer-Verlag in their lecture note series. At that conference, the organizers of the present conference asked me to give a similar series of two lectures here at Durham. For the sake of avoiding exact repetition, what I decided to do was to expand on two of the topics discussed there. In the first lecture, I discuss the existence of nilpotent elements in representation rings, treating this topic as an illustration of the above triangle. This may be regarded as a predigested version of Benson & Carlson [5]. In the second lecture, I go over the basic definitions and properties of the cohomological varieties associated to modular representations, and to illustrate the concepts I describe in some detail how to find the varieties for the indecomposable modules for the dihedral two-groups in characteristic two.

FIRST LECTURE

Nilpotent elements in representation rings

Throughout this lecture, G will be a finite group, and k an algebraically closed field of characteristic p. All modules will be finitely generated. We shall omit to mention that some of the results described here work with suitable modifications for more general rings of coefficients.

1. TENSOR PRODUCTS

Recall that the (modular) representation ring or Green Ring (after J.A. Green, who was the first to make any serious study of its structure) is the complex vector space $A(G)$ having the isomorphism classes $[V]$ of indecomposable kG-modules V as basis. Multiplication is given by $[V].[W] = [V \otimes W]$, where $V \otimes W$ denotes the tensor product over k, with the usual diagonal G-action. It is known that $A(G)$ is finite dimensional if and only if the Sylow p-subgroups of G are cyclic, and otherwise it is not even Noetherian. For most groups the indecomposable representations are in some sense unclassifiable (for groups with non-cyclic Sylow p-subgroups, the representation type is wild except when $p = 2$ and the Sylow 2-subgroups are dihederal, semidihedral, quaternion or generalized quaternion [8]).

Despite the fact that the tensor product operation has very widespread use in representation theory, in general very little is known about how a tensor product of indecomposable modules breaks up as a direct sum of indecomposables. This information is reflected in basic ring-theoretical properties of $A(G)$. As a first step, we determine when the trivial kG-module k appears as a direct summand of a tensor product. This appears to be one of the keys to understanding nilpotent elements in $A(G)$, as we shall see later in this lecture.

Theorem 1 ([5], Theorem 2.1) If V and W are indecomposable kG-modules then $V \otimes W$ has the trivial module k as a direct summand if and only if

(i) $V \cong W^*$

and (ii) $p \nmid \dim V$.

Moreover, under these conditions k is a summand with multiplicity one.

Sketch of proof The trivial kG-module k is a direct summand of $V \otimes W$ if and only if we can find homomorphisms $k \to V \otimes W$ and $V \otimes W \to k$ with non-zero composite. This happens if and only if the composite map

$$\mathrm{Hom}_{kG}(W^*,V) \xrightarrow{i} V \otimes W \xrightarrow{p} (\mathrm{Hom}_{kG}(V,W^*))^*$$

is non-zero. Associated to this we have a map

$$\mathrm{Hom}_{kG}(W^*,V) \otimes \mathrm{Hom}_{kG}(V,W^*) \xrightarrow{\eta} k$$

with $\eta \neq 0$ if and only if $p \circ i \neq 0$. It turns out that η is just composition followed by trace.

$$\mathrm{Hom}_{kG}(W^*,V) \otimes \mathrm{Hom}_{kG}(V,W^*) \xrightarrow{\circ} \mathrm{End}_{kG}(W^*) \xrightarrow{tr} k$$

Since W^* is indecomposable and k is algebraically closed, every endomorphism is of the form $\lambda I + n$ with n nilpotent. Since $tr(I) = \dim W^* = \dim W$, for η to be non-zero, we must have $p \nmid \dim W$, and we must have elements $\alpha \in \mathrm{Hom}_{kG}(W^*,V)$ and $\beta \in \mathrm{Hom}_{kG}(V,W^*)$ with $tr(\beta\circ\alpha) \neq 0$, namely such that $\beta \circ \alpha$ is an isomorphism. Since V is indecomposable this implies that $V \cong W^*$. The statement about multiplicities is not difficult.

2. ALMOST SPLIT SEQUENCES

Theorem 1 above may be translated into a statement about almost split sequences as follows. Recall from [6] or my Ottawa talks that with respect to the bilinear form (,) on $A(G)$ given by extending $([U], [V]) = \dim_k \mathrm{Hom}_{kG}(U,V)$ bilinearly, we have the following non-singularity statement. For each indecomposable module V we may find an element $\tau_0(V) \in A(G)$ such that for U indecomposable

$$([U], \tau_0(V)) = \begin{cases} 1 & \text{if } U \cong V \\ 0 & \text{otherwise.} \end{cases}$$

Namely

$$\tau_0(V) = \begin{cases} [V] - [\mathrm{Rad}\, V] & \text{if } V \text{ is projective} \\ [V] + [\Omega^2 V] - [X_V] & \text{otherwise} \end{cases}$$

where $0 \to \Omega^2 V \to X_V \to V \to 0$ is the almost split sequence terminating in V. In particular, it follows that if x is a non-zero element of $A(G)$, then there exists $y \in A(G)$ with $(x,y) \neq 0$.

With this in mind, Theorem 1 may now be interpreted as saying that for V and W indecomposable,

$$([V \otimes W^*], \tau_0(k)) = \begin{cases} 1 & \text{if } V \cong W \text{ and } p \nmid \dim V \\ 0 & \text{otherwise.} \end{cases}$$

Since $([V \otimes W^*], \tau_0(k)) = ([V], [W].\tau_0(k))$ this means that the following hold.

(i) If $p \mid \dim W$ then $([V], [W].\tau_0(k)) = 0$ for all indecomposable modules V. In particular taking $V = W$ this implies that the connecting homomorphism $\mathrm{Hom}_{kG}(W,W) \to \mathrm{Ext}^1_{kG}(W, \Omega^2 k \otimes W)$ is zero, and hence the sequence $0 \to \Omega^2 k \otimes W \to X_k \otimes W \to W \to 0$ splits.

(ii) If $p \nmid \dim W$ then for V indecomposable

$$([V], [W].\tau_0(k)) = \begin{cases} 1 & \text{if } V \cong W \\ 0 & \text{otherwise,} \end{cases}$$

and so the sequence $0 \to \Omega^2 k \otimes W \to X_k \otimes W \to W \to 0$ satisfies the defining conditions for an almost split sequence, apart from possibly the indecomposability of $\Omega^2 k \otimes W$. This means that if we strip off an injective (= projective) direct summand from the first two terms of this sequence, we are left with an almost split sequence.

We have thus outlined a proof of the following theorem. A different proof appears in Auslander & Carlson [2], which also gives the corresponding result for RG-lattices, where R is a complete discrete valuation ring.

<u>Theorem 2</u> ([2] Theorem 3.6; [5] Proposition 2.15) Let $0 \to \Omega^2 k \to X_k \to k \to 0$ be the almost split sequence terminating in the trivial module k. Let W be an indecomposable kG-module. Then the tensor product

$$0 \to \Omega^2 k \otimes W \to X_k \otimes W \to W \to 0$$

has the following properties.

(i) It is either split, or almost split modulo an injective summand.

(ii) It fails to split if and only if $p \nmid \dim W$.

3. NILPOTENT ELEMENTS IN A(G)

We now translate Theorem 1 into information about the structure of A(G). Let us denote by A(G;p) the linear span in A(G) of the indecomposable modules whose dimension is divisible by p.

Theorem 3 ([5], lemma 2.5 and theorem 2.7)

(i) A(G;p) is an ideal in A(G).

(ii) A(G)/A(G;p) has no non-zero nilpotent elements.

Sketch of proof (i) This statement is the same as the statement that if V is indecomposable and $p \mid \dim V$ then for any module W and any direct summand U of $V \otimes W$, we have $p \mid \dim U$. But this follows by applying theorem 1 to each side of the equality $(V \otimes W) \otimes U^* \cong V \otimes (W \otimes U^*)$.

(ii) First suppose $x = \Sigma a_i[V_i] \in A(G)$ with $xx^* \in A(G;p)$, where $x^* = \Sigma \bar{a}_i[V_i^*]$. Then $xx^* = \Sigma |a_i|^2 [V_i \otimes V_i^*] + \sum_{i \neq j} a_i \bar{a}_j [V_i \otimes V_j^*]$ does not involve the trivial module k, and so we may deduce from theorem 1 that $x \in A(G;p)$.

Now suppose $x \in A(G)$ with $x^2 \in A(G;p)$. Let $y = xx^*$. Then $yy^* \in A(G;p)$, and hence $y \in A(G;p)$, and hence $x \in A(G;p)$.

If H is a subgroup of G, we denote by $r_{G,H}$ the restriction map $A(G) \to A(H)$. It can be shown using an induction theorem of Conlon that if the Sylow p-subgroups of G are cyclic then $\bigcap_{H \leq G} r_{G,H}^{-1}(A(H;p)) = 0$. Thus we may deduce in this case that A(G) is semisimple. This was first proved by Green and O'Reilly in the sixties using some long-winded calculations. The above proof is due to Benson & Carlson [5], where more details may be found. To summarize, we have seen that whenever A(G) is finite dimensional it is semisimple.

4. COHOMOLOGICAL CONSTRUCTION OF NILPOTENT ELEMENTS

When A(G) is infinite dimensional the story is quite different. Zemanek [17,18] was the first to show that there can be nilpotent elements in A(G), by explicit construction of modules $V \not\cong W$ with $V \otimes V \oplus W \otimes W \cong V \otimes W \oplus V \otimes W$, so that $([V] - [W])^2 = 0$ in A(G). I shall outline a general construction due to Benson and Carlson [5], using cohomological techniques.

The construction depends on some modules L_ζ introduced by

Carlson [10] for the purpose of studying the cohomological varieties associated to modules. They now represent a standard method of passing from cohomology to representation theory. The definition is as follows. If $0 \neq \zeta \in H^{2n}(G,k) \cong \mathrm{Ext}^{2n}_{kG}(k,k) \cong \mathrm{Hom}_{kG}(\Omega^{2n}k,k)$ (where ΩV represents the kernel of the projective cover of V) then ζ is represented by a surjective homomorphism $\hat{\zeta} : \Omega^{2n}k \to k$, whose kernel we denote by L_ζ. If $\zeta = 0$, we make the convention $L_\zeta = \Omega^{2n}k \oplus \Omega k$. The basic lemma is as follows.

<u>Lemma</u> ([11]; [5], Theorem 3.3) Suppose V is a kG-module, and suppose $\zeta \in H^{2n}(G,k)$ annihilates $\mathrm{Ext}^*_{kG}(V,V)$ (cup-product action; note that it is enough to check that ζ annihilates the identity element in $\mathrm{Ext}^0_{kG}(V,V) = \mathrm{End}_{kG}(V)$). Then

$$L_\zeta \otimes V \cong \Omega^{2n}V \oplus \Omega V \oplus (\text{projective}).$$

Now suppose L_ζ happens to be periodic with period two (i.e. $\Omega^2 L_\zeta \cong L_\zeta$ but $\Omega L_\zeta \not\cong L_\zeta$), and suppose ζ annihilates $\mathrm{Ext}^*_{kG}(L_\zeta,L_\zeta)$. It happens that this forces G to have p-rank one or two, but that under these conditions there are many examples of such behaviour. We shall list some below. According to the lemma, we have

$$\begin{aligned} L_\zeta \otimes L_\zeta &\cong \Omega^{2n}L_\zeta \oplus \Omega L_\zeta \oplus (\text{projective}) \\ &\cong L_\zeta \oplus \Omega L_\zeta \oplus (\text{projective}). \end{aligned}$$

Applying Ω twice to this, we obtain

$$\Omega L_\zeta \otimes L_\zeta \cong \Omega L_\zeta \oplus L_\zeta \oplus (\text{projective})$$

and

$$\Omega L_\zeta \otimes \Omega L_\zeta \cong L_\zeta \oplus \Omega L_\zeta \oplus (\text{projective}).$$

Hence $([L_\zeta] - [\Omega L_\zeta])^2$ is a linear combination of projective modules. If L_ζ and ΩL_ζ have the same Brauer character (for a p-group this simply says that they have the same dimension) then $([L_\zeta] - [\Omega L_\zeta])^2 = 0$; otherwise we would have to adjust by some linear combination of projective modules to force the square to be zero (note that the linear span in $A(G)$ of the projective modules is a finite dimensional direct summand, so that we may project onto its complement).

We thus have the following theorem, which is a restricted version of [5], Theorem 3.4.

Theorem 4 Suppose $\zeta \in H^{2n}(G,k)$ has the following properties.

(i) L_ζ is periodic with period two.

(ii) L_ζ and ΩL_ζ have the same Brauer character.

(iii) ζ annihilates $\mathrm{Ext}^*_{kG}(L_\zeta, L_\zeta)$.

Then $[L_\zeta] - [\Omega L_\zeta]$ is a non-zero nilpotent element of $A(G)$.

Examples (a) For G an elementary abelian group of order p^2, p odd, and $k = \bar{\mathbb{F}}_p$, we have $H^*(G,k) = \Lambda(x_1,x_2) \otimes k[y_1,y_2]$ with $\deg x_1 = \deg x_2 = 1$ and $\deg y_1 = \deg y_2 = 2$. Then for each $(\alpha : \beta) \in \mathbb{P}^1(k)$ and each $n \geq 1$, the element $(\alpha y_1 + \beta y_2)^n \in H^{2n}(G,k)$ satisfies the conditions of theorem 4. The ideal in $A(G)$ generated by the corresponding nilpotent elements is an infinite dimensional nilpotent ideal. The same also works for $C_2 \times C_4$.

(b) For G dihedral of order 2^n, $n \geq 3$, and $k = \bar{\mathbb{F}}_2$, we have $H^*(G,k) = k[x,y,z]/(xy)$ with $\deg x = \deg y = 1$ and $\deg z = 2$. If $\alpha \neq 0 \neq \beta$ and $n \geq 1$ then the element $(\alpha x^2 + \beta y^2)^n \in H^{2n}(G,k)$ satisfies the conditions of theorem 4, and so again we get an infinite dimensional nilpotent ideal in $A(G)$.

(c) If G is a generalized quaternion group, the above construction produces nilpotent elements, but not in such vast quantities. The problem is that all kG-modules are periodic.

(d) If G is an elementary abelian 2-group, the above method produces no nilpotent elements at all, since every periodic module has period one. For G elementary abelian of order four, it turns out that there are no nilpotent elements (Conlon [12]). For G elementary abelian of order 2^n, $n \geq 3$, it is still an open problem as to whether $A(G)$ has nilpotent elements. Since there are standard methods for passing from subquotients to the whole group, this is essentially the only open case.

Remark The above nilpotent elements all square to zero. I do not know whether there are ever nilpotent elements of order greater than two, although this seems likely.

SECOND LECTURE

VARIETIES FOR MODULES : A SAMPLE CALCULATION

In this lecture I shall describe the theory of cohomological varieties associated to modular representations, and give the details of a sample calculation to illustrate the concepts. These varieties grew out of work of Quillen on the structure of equivariant cohomology rings [14,15], and have been developed by Carlson, Alperin, Evens, Avrunin, Scott [1,3,9,10,11] etc.

1. Definitions and basic properties

Let G be a finite group and k an algebraically closed field of characteristic p. Denote by $H^{\cdot}(G,k) \cong \mathrm{Ext}^{\cdot}_{kG}(k,k)$ the cohomology ring $H^{*}(G,k)$ if $p = 2$, and $H^{ev}(G,k)$, the even part of the cohomology ring if $p \neq 2$. Since the cohomology ring is graded commutative (i.e. $xy = (-1)^{\deg(x)\deg(y)}yx$), $H^{\cdot}(G,k)$ is a commutative ring, and so we may form the maximal ideal spectrum $X_G = \mathrm{Max}\, H^{\cdot}(G,k)$. Since the cohomology ring is finitely generated, we may view X_G as a concrete affine variety in the usual way. Namely if $H^{\cdot}(G,k) = k[x_1,\ldots,x_n]/I$ for some homogeneous generators x_i and homogeneous ideal I, then X_G is the variety in $\mathbb{A}^n(k)$ given by the simultaneous zeros of the polynomials in I. In particular, X_G is a homogeneous variety (a union of lines through the origin, where the origin corresponds to the ideal of elements of positive degree) and we may form a projective variety $\bar{X}_G = \mathrm{Proj}\, H^{\cdot}(G,k)$ of one smaller dimension.

Now if V is a KG-module, we may think of $\mathrm{Ext}^{*}_{kG}(V,V)$ as equivalence classes of long exact sequences beginning and ending in V, where the equivalence relation is generated by morphisms of long exact sequences which are isomorphisms on the end terms (see Maclane[13]). The Yoneda splice of long exact sequences gives $\mathrm{Ext}^{*}_{kG}(V,V)$ a ring structure, which may be non-commutative. In fact this ring may have complete matrix rings as quotients. However, there is a natural map $H^{\cdot}(G,k) \cong \mathrm{Ext}^{\cdot}_{kG}(k,k) \to \mathrm{Ext}^{*}_{kG}(V,V)$ given by tensoring long exact sequences with V. The image of this map lies in the centre of $\mathrm{Ext}^{*}_{kG}(V,V)$, and $\mathrm{Ext}^{*}_{kG}(V,V)$ is finitely generated as a module over the image. Thus if we let $\tilde{X}_G(V) = \mathrm{Max}\, Z\, \mathrm{Ext}^{*}_{kG}(V,V)$, the spectrum of maximal ideals of the centre of this ring, then we have a map of varieties $\tilde{X}_G(V) \to X_G$. We denote by $X_G(V)$ the image of this map, so that the map $\tilde{X}_G(V) \to X_G(V)$ is finite (the preimage of a point is a finite set). Also, since $X_G(V)$ is a (closed) homogeneous subvariety of X_G, we may form the corresponding projective subvariety $\bar{X}_G(V)$ of $\bar{X}_G$.

<u>Example</u> If G is elementary abelian of order p^n, then

$$H^*(G,k) = \begin{cases} k[x_1,\dots,x_n] & \text{if } p = 2 \\ \Lambda(x_1,\dots,x_n) \otimes k[y_1,\dots,y_n] & \text{if } p \neq 2 \end{cases}$$

where $\deg(x_i) = 1$, $\deg(y_i) = 2$, and $\Lambda(x_1,\dots,x_n)$ denotes the exterior algebra. Thus if we denote by J the radical of nilpotent elements in $H^{\cdot}(G,k)$ then $H^{\cdot}(G,k)/J$ is a polynomial ring in n variables (x_i if $p = 2$ and y_i if $p \neq 2$) and so $X_G \cong \mathbb{A}^n(k)$, $\bar{X}_G \cong \mathbb{P}^{n-1}(k)$.

For more general G, Quillen [14,15] has shown that

$$X_G \cong \varinjlim_{E \leq G} X_E.$$

The direct limit is taken over the set of all elementary abelian subgroups E of G, and the maps are those induced by conjugations and inclusions. The "isomorphism " is a homeomorphism in the Zariski topology, or an "inseparable isogeny" at the coordinate ring level. Avrunin and Scott [3] have shown that the appropriate generalization to modules is also true:

$$X_G(V) = \varinjlim_{E \leq G} X_E(V{\downarrow_E}).$$

From the definitions and results given so far, it would seem that the varieties $X_G(V)$ are very difficult to calculate. It turns out that there is an alternative formulation which makes calculation much easier. By the above result of Avrunin and Scott, it suffices to treat the case where G is elementary abelian. In this case, we have the following <u>rank variety</u> introduced by Carlson [10].

Let Y_G be the affine space $J(kG)/J^2(kG) \cong \mathbb{A}^n$ (where $|G| = p^n$), and let $Y_G(V)$ denote the image modulo J^2 of $\{0\} \cup \{\alpha \in J(kG) : V{\downarrow_{\langle 1+\alpha \rangle}}$ is not a free $k\langle 1+\alpha \rangle$ - module$\}$ (it turns out that this is a union of cosets of J^2). There is a natural isomorphism

$X_G \cong Y_G$ with the property that for all kG-modules V, the image of $X_G(V)$ is equal to $Y_G(V)$. This was conjectured by Carlson and proved by Avrunin and Scott, as the difficult step in proving their theorem mentioned above. The proof of statement (ii) of Theorem 1 below also needed this fact. The variety $Y_G(V)$ is easy to calculate as the set of zeros of minors of certain matrices.

The following is a list of properties of the $X_G(V)$ which we shall be using. Some of these properties are quite difficult to prove.

Theorem 1 Let V and W be kG-modules.

(i) $X_G(V \oplus W) = X_G(V) \cup X_G(W)$

(ii) (Avrunin-Scott [3]) $X_G(V \otimes W) = X_G(V) \cap X_G(W)$

(iii) The dimension of the variety $X_G(V)$ is equal to the complexity of V. Namely if $\ldots \to P_2 \to P_1 \to P_0 \to V \to 0$ is a minimal resolution of V, the complexity is defined to be the order of the pole of the rational function $\Sigma t^i \dim P_i$ at $t = 1$. This measures the rate of growth of the resolution. Thus the complexity is zero if and only if V is projective, and one if and only if V is periodic, and so on.

(iv) (Carlson [11]) If V is indecomposable then $\bar{X}_G(V)$ is connected in the Zariski topology.

(v) (Carlson [10]) If $0 \neq \zeta \in H^{2n}(G,k) \cong \mathrm{Ext}^{2n}_{kG}(k,k) \cong \mathrm{Hom}_{kG}(\Omega^{2n}k,k)$, we define L_ζ to be the kernel of the homomorphism $\Omega^{2n}k \to k$ (as in the last lecture). Then $X_G(L_\zeta)$ is the hypersurface $X_G(\zeta)$ given by taking the zeros of ζ regarded as an element of the coordinate ring of X_G.

Proofs of these statements may be found in [1,3,9,10,11]. It should be remarked that it follows from (ii) and (v) of this theorem that every homogeneous (closed) subvariety of X_G is of the form $X_G(V)$ (express the variety as an intersection of hypersurfaces, and take the tensor product of the corresponding L_ζ's).

We now turn to our sample calculation to illustrate the above concepts. We have chosen a class of groups for which we have a good understanding of the set of indecomposable modules, namely the dihedral two-groups. The results of these calculations appear without proof in the appendix of [4], p.185.

2. Cohomology of the dihedral two-groups

For the rest of this lecture, let $G = \langle u,v : u^2 = v^2 = (uv)^{2^{n-1}} = 1\rangle$ be the dihedral group of order 2^n, and let k be an algebraically closed field of characteristic two. Then $H^*(G,k) = k[x,y,z]/(xy)$, where $\deg(x) = \deg(y) = 1$ and $\deg(z) = 2$. We choose the labelling in such a way that the generator $x \in H^1(G, \mathbb{F}_2) = \mathrm{Hom}(G, \mathbb{Z}/2\mathbb{Z})$ corresponds to the subgroup $\langle uvu,v\rangle$ of index two, while y corresponds to $\langle u,vuv\rangle$. Thus $\bar{X}_G = \mathrm{Proj}(k[x,y,z]/(xy)) = \mathbb{P}^1_a \cup \mathbb{P}^1_b$, where $\mathbb{P}^1_a$ and $\mathbb{P}^1_b$ are projective lines over k intersecting in the common point at infinity : $\mathbb{P}^1_a \cap \mathbb{P}^1_b = \{\infty_a = \infty_b\}$. We choose the notation so that $\mathbb{P}^1_a = \mathrm{Proj}(k[x,z])$ and $\mathbb{P}^1_b = \mathrm{Proj}(k[y,z])$, and so that $\lambda x^2 + \mu y^2 + z = 0$ is the equation of the pair of points $\{\lambda_a,\mu_b\} \subseteq (\mathbb{P}^1_a \cup \mathbb{P}^1_b)\backslash\{\infty\}$.

Let $H_a = \langle u,w\rangle$ and $H_b = \langle v,w\rangle$ where $w = (uv)^q$, $q = 2^{n-2}$, be representatives of the two conjugacy classes of elementary abelian subgroups of order four in G. Then $H^*(H_a,k) = k[x_a,z_a]$ and $H^*(H_b,k) = k[y_b,z_b]$ with $\deg(x_a) = \deg(z_a) = \deg(y_b) = \deg(z_b) = 1$. We choose the notation so that $x_a = \mathrm{res}_{G,H_a}(x)$, $y_b = \mathrm{res}_{G,H_b}(y)$, z_a corresponds to the subgroup $\langle u\rangle$ of index two in H_a, and z_b corresponds to the subgroup $\langle v\rangle$ in H_b. Then the restriction maps are as follows.

Generator of $H^*(G,k)$	Image under res_{G,H_a}	Image under res_{G,H_b}
x	x_a	0
y	0	y_b
z	$z_a(x_a + z_a)$	$z_b(y_b + z_b)$

Let $\hat{\mathbb{P}}^1_a = \mathrm{Proj}\, H^*(H_a,k)$, labelled in such a way that u corresponds to $\hat{0}_a \in \hat{\mathbb{P}}^1_a$, w corresponds to $\hat{\infty}_a$, and uw corresponds to $\hat{1}_a$. Similarly we label $\hat{\mathbb{P}}^1_b = \mathrm{Proj}\, H^*(H_b,k)$ in such a way that v corresponds to $\hat{0}_b \in \hat{\mathbb{P}}^1_b$, w corresponds to $\hat{\infty}_b$, and vw corresponds to $\hat{1}_b$. The maps $t_{H_a,G} = \mathrm{res}^*_{G,H_a} : \hat{\mathbb{P}}^1_a \to \mathbb{P}^1_a$ and $t_{H_b,G} = \mathrm{res}^*_{G,H_b} : \hat{\mathbb{P}}^1_b \to \mathbb{P}^1_b$ are given by

$$t_{H_a,G}(\hat{\lambda}_a) = \lambda_a(1+\lambda_a)$$

$$t_{H_b,G}(\hat{\mu}_b) = \mu_b(1+\mu_b).$$

3. MODULES FOR THE DIHEDRAL TWO-GROUPS

The indecomposable kG-modules $(G = D_{2^n})$ were first classified by Bondarenko [7], but we shall rather use the description given in Ringel [16]. First we describe the finite dimensional indecomposable modules for the infinite dihedral group

$$\tilde{G} = \langle u,v : u^2 = v^2 = 1\rangle$$

and then we indicate which are modules for the quotient group $G = D_{2^n} = \langle u,v : u^2 = v^2 = 1, (uv)^q = (vu)^q\rangle$.

Let $\mathcal{W}$ be the set of words in the letters a, b, a^{-1} and b^{-1} such that a and a^{-1} are always followed by b or b^{-1} and vice-versa, together with the "zero length words" 1_a and 1_b. If C is a word, we define C^{-1} as follows. $(1_a)^{-1} = 1_b$, $(1_b)^{-1} = 1_a$; and otherwise, we reverse the order of the letters in the word and invert each letter according to the rule $(a^{-1})^{-1} = a$, $(b^{-1})^{-1} = b$. Let $\mathcal{W}_1$ be the set obtained from $\mathcal{W}$ by identifying each word with its inverse.

The nth power of a word of even length is obtained by juxtaposing n copies of the word. Let $\mathcal{W}'$ be the subset of $\mathcal{W}$ consisting of all words of even non-zero length which are not powers of smaller words. Let $\mathcal{W}_2$ be the set obtained from $\mathcal{W}'$ by identifying each word with its inverse and with its images under cyclic permutations of the letters, $\ell_1 \ldots \ell_n \to \ell_n \ell_1 \ldots \ell_{n-1}$.

The following is a list of all the isomorphism types of indecomposable $k\tilde{G}$-modules.

<u>Modules of the first kind</u> These are in one-one correspondence with elements of $\mathcal{W}_1$. Let $C = \ell_1 \ldots \ell_n \in \mathcal{W}$. Let $M(C)$ be a vector space over k with basis $z_0, \ldots, z_n$ on which $\tilde{G}$ acts according to the schema

$$kz_0 \xleftarrow{\ell_1} kz_1 \xleftarrow{\ell_2} kz_2 \ldots kz_{n-1} \xleftarrow{\ell_n} kz_n$$

where x acts as "1 + a" and y acts as "1 + b". For example, if $C = ab^{-1}aba^{-1}$ then the schema is

$$kz_0 \xleftarrow{a} kz_1 \xrightarrow{b} kz_2 \xleftarrow{a} kz_3 \xleftarrow{b} kz_4 \xrightarrow{a} kz_5$$

and the representation is given by

$$x \mapsto \begin{pmatrix} 1 & & & & & \\ 1 & 1 & & & & \\ & & 1 & & & \\ & & 1 & 1 & & \\ & & & & 1 & 1 \\ & & & & & 1 \end{pmatrix} \qquad y \mapsto \begin{pmatrix} 1 & & & & & \\ & 1 & 1 & & & \\ & & 1 & & & \\ & & & 1 & & \\ & & & 1 & 1 & \\ & & & & & 1 \end{pmatrix}$$

It is clear that $M(C) \cong M(C^{-1})$.

<u>Modules of the second kind</u>. These are in one-one correspondence with elements of $\mathcal{W}_2 \times \mathcal{V}$ where

$$\mathcal{V} = \{(V,\phi) : V \text{ is a vector space over } k \text{ and } \phi \text{ is an indecomposable automorphism of } V\}$$

(since we are only dealing with the case where k is algebraically closed, an indecomposable automorphism of a vector space is simply a Jordan block). If $(C,(V,\phi)) \in \mathcal{W}' \times \mathcal{V}$ with $C = \ell_1 \ldots \ell_n$, let $M(C,V,\phi)$ be the vector space $\bigoplus_{i=0}^{n-1} V_i$ with $V_i \cong V$ on which G acts according to the schema

$$V_0 \xleftarrow{\ell_1=\phi} V_1 \xleftarrow{\ell_2=\mathrm{id}} V_2 \;\cdot\;\cdot\; V_{n-2} \xleftarrow{\ell_{n-1}=\mathrm{id}} V_{n-1}, \qquad V_0 \xrightarrow{\ell_n=\mathrm{id}} V_{n-1}$$

where again x acts as "1 + a" and y acts as "1 + b" as above. It is clear that if C and C' represent the same element of $\mathcal{W}_2$ then $M(C,V,\phi) \cong M(C',V,\phi)$.

A module represents the quotient group G if and only if either

(i) the module is of the first kind and the corresponding word does not contain $(ab)^q$, $(ba)^q$ or their inverses,

(ii) the module is of the second kind and no power of the corresponding word contains $(ab)^q$, $(ba)^q$ or their inverses, or

(iii) the module is the projective indecomposable module $M((ab)^q(ba)^{-q},k,id)$ (of the second kind).

Ringel [16] also calculated which of the above modules are periodic. It turns out that a module of the first kind $M(C)$ is periodic if and only if $C \sim (ab)^{q-1}a$ or $(ba)^{q-1}b$, while all modules of the second kind are periodic.

4. THE VARIETIES FOR THE INDECOMPOSABLE kD_{2n}-MODULES

The following theorem gives the varieties for the indecomposable kG-modules in terms of the above classification.

Theorem 2 (i)

$$\bar{X}_G(M(C)) = \begin{cases} \mathbb{P}^1_a \cup \mathbb{P}^1_b & \text{if } C \sim a^{\pm 1} \ldots b^{\pm 1} \\ \mathbb{P}^1_b & \text{if } C \sim a^{\pm 1} \ldots a^{\pm 1} \\ & \quad \text{but } C \not\sim (ab)^{q-1}a \\ \mathbb{P}^1_a & \text{if } C \sim b^{\pm 1} \ldots b^{\pm 1} \\ & \quad \text{but } C \not\sim (ba)^{q-1}b \\ \{O_b\} & \text{if } C \sim (ab)^{q-1}a \\ \{O_a\} & \text{if } C \sim (ba)^{q-1}b \end{cases}$$

(ii)

$$\bar{X}_G\left(M\left(C, \begin{pmatrix} \lambda & 1 & & 0 \\ & \ddots & \ddots & \\ & & \ddots & 1 \\ 0 & & & \lambda \end{pmatrix}\right)\right) = \begin{cases} \{\infty\} & \text{unless } C \sim a^{-1}b(ab)^{q-1}, \\ & \quad b^{-1}a(ba)^{q-1} \text{ or } (ab)^q(ba)^{-q} \\ \lambda_a & \text{if } C \sim a^{-1}b(ab)^{q-1} \\ \lambda_b & \text{if } C \sim b^{-1}a(ba)^{q-1} \end{cases}$$

(iii) $\bar{X}_G(M((ab)^q(ba)^{-q},k,id)) = \phi$.

We shall prove Theorem 2 by dealing with the various cases in separate lemmas. The following lemma deals with the first case of (i).

Lemma 1. If V is an indecomposable kG-module with $\dim(V)$ odd, then $\bar{X}_G(V) = \bar{X}_G$.

Proof. If $\dim(V)$ is odd, then for each shifted subgroup $\langle 1+\alpha \rangle$ of an elementary abelian subgroup E of G, $V{\downarrow}_{\langle 1+\alpha\rangle}$ is not free. Thus $Y_E(V{\downarrow}_E) = Y_E$, hence $X_E(V{\downarrow}_E) = X_E$, and so $X_G(V) = \varinjlim_{E} X_E(V{\downarrow}_E) = X_G$.

Lemma 2. A kG-module $M(C)$ of the first kind is periodic if and only if $C \sim (ab)^{q-1}a$ or $C \sim (ba)^{q-1}b$.

Proof. As mentioned at the end of section 3, this was proved by Ringel in [16].

Lemma 3. (i) $M(a^{\pm 1} \dots a^{\pm 1}){\downarrow}_{\langle u\rangle}$ is free while $M(a^{\pm 1} \dots a^{\pm 1}){\downarrow}_{\langle v\rangle}$ is not.

(ii) $M(b^{\pm 1} \dots b^{\pm 1}){\downarrow}_{\langle v\rangle}$ is free while $M(b^{\pm 1} \dots b^{\pm 1}){\downarrow}_{\langle u\rangle}$ is not.

Proof. This follows from the explicit description of the action of u and v on these modules given by the schema. Thus it can be seen, for example, that $M(a^{\pm 1} \dots a^{\pm 1}){\downarrow}_{\langle v\rangle}$ has exactly two non-projective summands, corresponding to the basis elements occuring at the end of the schema.

Lemma 4. (i) $$\bar{X}_G(M(a^{\pm 1} \dots a^{\pm 1})) = \begin{cases} \{o_b\} & \text{if } C \sim (ab)^{q-1}a \\ \mathbb{P}^1_b & \text{otherwise} \end{cases}$$

(ii) $$\bar{X}_G(M(b^{\pm 1} \dots b^{\pm 1})) = \begin{cases} \{o_a\} & \text{if } C \sim (ba)^{q-1}b \\ \mathbb{P}^1_a & \text{otherwise.} \end{cases}$$

Proof. Carlson's connectedness theorem (part (iv) of theorem 1) states that if V is indecomposable then $\bar{X}_G(V)$ is connected in the Zariski topology. As explained in section 1, we may calculate $\bar{X}_G(V)$ by restrictions to cyclic subgroups $\langle 1+\alpha\rangle$ of kG. Thus it follows that $\bar{X}_G(M(a^{\pm 1} \dots a^{\pm 1}))$ is a connected subvariety of $\bar{X}_G$ containing the point o_b but not o_a, by Lemma 3. If $C \sim (ab)^{q-1}a$ then by Lemma 2, $M(C)$ is periodic, and so by part (iii) of Theorem 1, $\bar{X}_G(M(C)) = \{o_b\}$. For all other choices of $C = a^{\pm 1} \dots a^{\pm 1}$, $M(C)$ is not periodic, and so $\dim(\bar{X}_G(M(C))) = 1$. Thus $\bar{X}_G(M(C)) = \mathbb{P}^1_b$. Statement (ii) is proved in the same way.

We have now completed the proof of part (i) of Theorem 1, and so we turn to the non-projective modules of the second kind. According to section 8 of [16], these are all periodic of period 1 or 2, and so by Carlson's connectedness theorem their variety consists of a single point in each case.

Lemma 5. Suppose V is a non-projective indecomposable kG-module $M(C,\phi)$ of the second kind. Then $\bar{X}_G(V) = \{\infty\}$ unless $C \sim a^{-1}b(ab)^{q-1}$ or $C \sim b^{-1}a(ba)^{q-1}$, in which cases we have $\bar{X}_G(V) \subseteq \mathbb{P}^1_a\setminus\{\infty\}$, resp. $\bar{X}_G(V) \subseteq \mathbb{P}^1_b\setminus\{\infty\}$.

Proof. Suppose $\bar{X}_G(V) \neq \{\infty\}$. By the above remark, $\bar{X}_G(V)$ is either a point in $\mathbb{P}^1_a\setminus\{\infty\}$ or a point in $\mathbb{P}^1_b\setminus\{\infty\}$. Suppose without loss of generality that we are in the former case. Then $\bar{X}_{H_b}(V\downarrow_{H_b}) = \phi$, and so $\bar{X}_H(V\downarrow_H) = \phi$, where $H = \langle uvu,v\rangle$ of index two in G. Hence by part (iii) of Theorem 1, $V\downarrow_H$ is projective.

Now when dealing with modules for a p-group, we can distinguish projective modules from non-projective modules by the action of the norm element (i.e. the sum of all the group elements as an element of the group algebra). For the norm element acts as zero on all non-projective indecomposable modules. Thus the rank of the corresponding matrix in a given representation is at most the dimension divided by the order of the group, with equality if and only if the representation is projective.

The norm element of H is

$$n_H = [(1+uvu)(1+v)]^{q/2}.$$

Let $X = 1+u$ and $Y = 1+v$, so that $X^2 = Y^2 = 0$. Then

$$\begin{aligned} n_H &= [(Y + XY + YX + XYX)Y]^{q/2} \\ &= (YXY + XYXY)^{q/2} \\ &= (YX)^{q-1}Y + (XY)^q. \end{aligned}$$

Since V is a non-projective indecomposable kG-module, $(XY)^q$, which is the norm element of kG, acts as zero. Since $V\downarrow_H$ is a projective kH-module, n_H, which we now know to act in the same way as $(YX)^{q-1}Y$, acts as a matrix whose rank is $\dim(V)/|H|$. So the rank of $(YX)^{q-1}Y$ on V is dim(V)/2q. Looking at the description of how X and Y act on V according to the schema given in section 3, we see that this is impossible unless we have $C \sim [a^{-1}b(ab)^{q-1}]^r$ for some $r \geq 1$ (recall that $(ba)^q$ must not appear in any power of C). Now since a word in $\mathcal{W}'$ is not allowed to be a power of a word of smaller length, we have $r = 1$.

A similar argument shows that in the case where $\bar{X}_G(V)$ is a point in $\mathbb{P}^1_b \backslash \{\infty\}$, then $C \sim b^{-1}a(ba)^{q-1}$.

To complete the proof of Theorem 2, we must identify some modules of the form L_ζ, and use part (v) of Theorem 1.

<u>Lemma 6</u> Let $\zeta = (\lambda x^2 + \mu y^2 + z)^r \in H^{2r}(G,k)$. Then

$$L_\zeta \cong M(a^{-1}b(ab)^{q-1}, \begin{pmatrix} \lambda & 1 & & 0 \\ & \cdot & \cdot & \\ & & \cdot & \cdot \\ & & & \cdot\ 1 \\ 0 & & & \lambda \end{pmatrix}) \oplus M(b^{-1}a(ba)^{q-1}, \begin{pmatrix} \mu & 1 & & 0 \\ & \cdot & \cdot & \\ & & \cdot & \cdot \\ & & & \cdot\ 1 \\ 0 & & & \mu \end{pmatrix}),$$

where the matrices on the right hand side of this equation are $r \times r$ matrices.

<u>Proof.</u> It is easy to see by direct calculation or by looking at section 8 of [16] that $\Omega^{2r}(k) \cong M((b^{-1}a(ba)^{q-1})^{-r}(a^{-1}b(ab)^{q-1})^r)$, a module of dimension $4qr+1$. According to the schema, we have an ordered basis $z_0, \dots, z_{4qr}$ corresponding to this word.

With respect to this basis, $y^{2(r-s)}z^s \in H^{2r}(G,k)$ corresponds to the homomorphism from $\Omega^{2r}(k)$ to k sending z_{2qs} to 1 and all other z_i to zero, while $x^{2(r-s)}z^s$ corresponds to the homomorphism sending $z_{2q(2r-s)}$ to 1 and all other z_i to zero. We shall show that

$$L_\zeta \cap \langle z_0, \dots, z_{2qr}\rangle \cong M(b^{-1}a(ba)^{q-1}, \begin{pmatrix} \mu & 1 & & 0 \\ & \cdot & \cdot & \\ & & \cdot & \cdot \\ & & & \cdot\ 1 \\ 0 & & & \mu \end{pmatrix}), \text{ while}$$

$$L_\zeta \cap \langle z_{2qr}, \dots, z_{4qr}\rangle \cong M(a^{-1}b(ab)^{q-1}, \begin{pmatrix} \lambda & 1 & & 0 \\ & \cdot & \cdot & \\ & & \cdot & \cdot \\ & & & \cdot\ 1 \\ 0 & & & \lambda \end{pmatrix}).$$

In fact, we shall only show the latter, since the former is symmetrically identical. In terms of the schema for

$$M(a^{-1}b(ab)^{q-1}, \begin{pmatrix} \lambda & 1 & & 0 \\ & \ddots & \ddots & \\ & & \ddots & 1 \\ 0 & & & \lambda \end{pmatrix}), \text{ we wish to take}$$

$V_i = \langle z_{2qr+i}, z_{2q(r+1)+i}, \ldots, z_{2q(2r-1)+i} \rangle$ for $1 \leq i \leq 2q-1$, and
$V_0 = L_\zeta \cap \langle z_{2qr}, z_{2q(r+1)}, \ldots, z_{4qr} \rangle$, taking as basis for V_0 the images of $z_{2q(r+1)}, \ldots, z_{4qr}$ under the map $L_\zeta \cap \langle z_{2qr}, z_{2q(r+1)}, \ldots, z_{4qr} \rangle \overset{\cong}{\to} \langle z_{2qr} \rangle + (L_\zeta \cap \langle z_{2qr}, \ldots, z_{4qr} \rangle)/\langle z_{2qr} \rangle$. In terms of these bases, the map $\phi : V_1 \to V_0$ of the schema has as its matrix

$$\phi = \begin{pmatrix} r\lambda & \binom{r}{2}\lambda^2 & \ldots & \lambda^r \\ 1 & 0 & \ldots & 0 \\ 0 & 1 & \ldots & 0 \\ \vdots & & & \vdots \\ 0 & 0 & \ldots 1 & 0 \end{pmatrix}$$

Some elementary linear algebra shows that this is conjugate to the matrix

$$\begin{pmatrix} \lambda & 1 & & 0 \\ & \ddots & \ddots & \\ & & \ddots & 1 \\ 0 & & & \lambda \end{pmatrix}.$$

Thus we have produced two submodules of L_ζ of the appropriate isomorphism types, which intersect in $\{0\}$ and span L_ζ. This completes the proof.

We may now complete the proof of theorem 2. It follows from lemma 6 and part (v) of Theorem 1 that

$$\bar{X}_G(M(a^{-1}b(ab)^{q-1}, \begin{pmatrix} \lambda & 1 & & 0 \\ & \ddots & \ddots & \\ & & \ddots & 1 \\ 0 & & & \lambda \end{pmatrix}) \oplus M(b^{-1}a(ba)^{q-1}, \begin{pmatrix} \mu & 1 & & 0 \\ & \ddots & \ddots & \\ & & \ddots & 1 \\ 0 & & & \mu \end{pmatrix}))$$

$= \{\lambda_a, \mu_b\}$. Comparing these statements for different values of λ and μ, and using part (i) of Theorem 1, we see that part (ii) of Theorem 2 holds. Part (iii) follows from part (iii) of Theorem (i).

Open Problem Calculate the ring structure of $A(D_{2^n})$. It would be interesting to understand the nilpotent elements in this ring.

REFERENCES

[1] Alperin, J.L. & Evens, L. (1982). Varieties and elementary abelian subgroups. J. Pure Appl. Alg., 26, 221-227.

[2] Auslander, M. & Carlson, J.F. (to appear). Almost split sequences and group rings. J. Alg.

[3] Avrunin, G.S. & Scott, L.L. (1982). Quillen stratification for modules. Invent. Math. 66, 277-286.

[4] Benson, D.J. (1984). Modular representation theory : new trends and methods. Springer Lecture Notes in Mathematics no.1081, Berlin/New York, Springer-Verlag.

[5] Benson, D.J. & Carlson, J.F. (to appear). Nilpotent elements in the Green ring. J. Alg.

[6] Benson, D.J. & Parker, R.A. (1984). The Green ring of a finite group. J. Alg. 87, 290-331.

[7] Bondarenko, V.M. (1975). Representations of dihedral groups over a field of characteristic 2. Math. U.S.S.R. Sbornik, Vol.25,no.1.

[8] Bondarenko, V.M. & Drozd, Yu. A. (1977). The representation type of finite groups. Zap. Naučn. Sem. LOMI 57, 24-41.

[9] Carlson, J.F. (1981). Complexity and Krull dimension. In Representations of algebras. Springer Lecture Notes in Mathematics no. 903, p.62-67, Berlin/New York, Springer-Verlag.

[10] Carlson, J.F. (1983). Varieties and the cohomology ring of a module. J. Alg. 85, 104-143.

[11] Carlson, J.F. (1984). The variety of an indecomposable module is connected. Invent. Math. 77, 291-299.

[12] Conlon, S.B. (1965). Certain representation algebras. J. Austr. Math. Soc. 5, 83-99.

[13] MacLane, S. (1963). Homology, Berlin/New York, Springer-Verlag.

[14] Quillen, D. (1971). The spectrum of an equivariant cohomology ring, I. Ann. of Math. 94, 549-572.

[15] Quillen, D. (1971). The spectrum of an equivariant cohomology ring, II, Ann of Math. 94, 573-602.

[16] Ringel, C.M. (1975). The indecomposable representations of the dihedral 2-groups. Math. Ann. 214, 19-34.

[17] Zemanek, J.R. (1971). Nilpotent elements in representation rings. J. Alg. 19, 453-469.

[18] Zemanek, J.R. (1973). Nilpotent elements in representation rings over fields of characteristic 2. J. Alg. 25, 534-553.